GREEN SCREEN

Green Screen identifies the various ways in which the natural world and the built environment have been conceptualised in American culture, and analyses the interplay of environmental ideologies at work in Hollywood movies. David Ingram argues that Hollywood cinema plays an important ideological role in the 'greenwashing' of ecological discourses, while largely perpetuating romantic attitudes to nature, including those prevalent in deep ecological thought. These arcadian constructions remain ultimately at the service of a mainstream environmentalist agenda.

In classifying films as 'environmentalist', ***Green Screen*** does not presuppose that they treat their subject matter in a way that is either serious, complex or profound or that they present a single, coherent or clear intellectual position towards environmental issues. Rather, the central thesis of the book is that Hollywood environmentalist movies bring together a range of contradictory discourses concerning the relationship between human beings and the environment. The natural world is shown to be implicated in complex human conflicts over gender, ethnicity, class and national identity.

> 'This book is primarily an agenda-setter. As such it makes clear how complex and important are the debates that film studies and, more widely, American studies will need to tackle regarding representations and critique of late-capitalist consumerism in its global phase.'
>
> ***Forum for Modern Languages***

> '*Green Screen* combines film criticism, cultural criticism, ecocriticism, and a bit of environmental history in an engaging and useful way. Its selection of films, many of which are described in some detail, will be useful to those who are entering the field. Its insights will be of value to ecocritical scholars and to those who want to bring environmental film into their classroom.'
>
> ***Interdisciplinary Studies for Literature and the Environment***

David Ingram is a lecturer in the School of Arts at Brunel University.

REPRESENTING AMERICAN CULTURE
A series from University of Exeter Press
Series Editor: Mick Gidley

Narratives and Spaces: Technology and the Construction of American Culture
David E. Nye (1997)

The Radiant Hour: Versions of Youth in American Culture
edited by Neil Campbell (2000)

GREEN SCREEN

Environmentalism and Hollywood Cinema

David Ingram

UNIVERSITY
of
EXETER
PRESS

For Pam

First published in 2000 by
University of Exeter Press
Reed Hall, Streatham Drive
Exeter, Devon EX4 4QR
UK

www.exeterpress.co.uk

Reprinted 2002
First paperback edition 2004
Reprinted 2007, 2008

Printed digitally since 2010

British Library Cataloguing in Publication Data
A catalogue record of this book is available
from the British Library.

Hardback ISBN 978 0 85989 608 5
Paperback ISBN 978 0 85989 609 2

Typeset in 10/12.5 Caslon
by Quince Typesetting, Exeter

Cover photograph
Male and female killer whale *Orcinus orca*, Cork, Ireland, June 2001
(© Martin Elcoate)

Contents

Acknowledgements

I would like to thank the staff at the Library of Congress, Washington DC; the Margaret Herrick Library of the Academy of Motion Pictures, Arts and Sciences, Los Angeles; the Film Archive of the University of California at Los Angeles; the Film Center at the Museum of Modern Art, New York; and the British Film Institute and British Library in London, where the research for this project was mainly undertaken. I would especially like to thank Madeline Matz and her colleagues at the Library of Congress for arranging many happy hours of film viewing.

Thanks also to Professor Mick Gidley for initiating the project, Brunel University for financial support, the anonymous reader for his helpful comments and advice, the University of Exeter Press for bringing it to fruition, and to all of my colleagues in American Studies at Brunel University for their encouragement and good humour.

I would also like to thank my parents, Roy and Eileen, and, leaving the best until last, Pam, for her patience, understanding and support.

Chapter Five is an expanded and revised version of 'The Camera in the Garden: The Ecology and Politics of the Amazonian Rainforest in Hollywood Cinema', published in *Borderlines*, 3.4 (1996), 376–388, and is used with permission.

Preface

In September 1990, the *Hollywood Reporter* announced the arrival of a new movie genre: 'film vert'. When *Audubon* magazine confirmed 'the greening of Tinseltown' in March 1992, the 'green' movie, it seemed, had become an identifiable cycle within Hollywood film production.[1] The new trend centred mainly on the issue of rain forest depletion, which formed the premise for a varied group of films that included the comedy *Meet the Applegates* (1990), the drama *At Play in the Fields of the Lord* (1991), and the children's animation *FernGully: The Last Rainforest* (1992). It is with such self-consciously environmentalist films that this book is mainly concerned, tracing their history as far back as the silent era, when two film versions of Peter B. Kyne's novel *The Valley of the Giants* (1918) reconciled a desire to preserve a valley of giant sequoias for its spiritual value with the official conservationist ideologies of their day.

The environmental issues that inform the narrative of such films clearly occur as matters of degree. What this book refers to as an 'environmentalist' film, then, is a work in which an environmental issue is raised explicitly and is central to the narrative. Such a film is at one end of a continuum that includes, at the other, the vast majority of films in which representations of the environment serve merely as a background to the central human drama. However, the omission or denial of an environmentalist discourse in such films can itself be significant. Geographer Neil Smith comments that non-human nature is usually rendered in literature as 'a backdrop, a mood setter, at best a refractory image of, or rather simplistic metaphor for, specific human emotions and dramas that inscribe the text. The play of human passions is the thing'.[2] The same, of course, holds true for Hollywood cinema. Nevertheless, critical analysis of films of this type can bring to the foreground their unacknowledged, unreflective references to non-human nature, so that their environmentalist implications can become both visible and open to question.

In classifying films as 'environmentalist', *Green Screen* does not presuppose that they treat their subject matter in a way that is either serious, complex or profound. Nor does it imply that they present a single, coherent or clear intellectual position towards environmental issues. Indeed, many of the movies discussed in this book use such issues as a premise for the exploration of more familiar Hollywood concerns, in particular the testing of the white male hero in gender and ethnic relationships. In this sense, Hollywood cinema has treated environmentalism in the same way as all other topical issues. The institutional and ideological constraints of what Richard Maltby and Ian Craven call Hollywood's 'commercial aesthetic' have always placed a value on the pleasures of entertainment rather than on polemic. Political subjects are therefore appropriate when they can provide scriptwriters and directors with the human interest stories, 'dramatic potentials' and 'angles' that they require to make a commercial movie. Given the commercial imperatives of the industry, Maltby and Craven argue, there is no incentive for such movies to be politically clear. Instead, the representation of political issues tends to take the form of what they describe as 'exclusions, hesitations, and absences'.[3] Stephen Prince also argues that Hollywood's commercial intent to maximize profits by appealing to wide and diverse audiences works against ideological and political coherence in the films themselves. Instead, a Hollywood movie is typically what he calls an 'ideological agglomeration' that constructs a 'polysemous, multivalent set of images, characters, and narrative situations'.[4] The central thesis of this book, then, is that Hollywood environmentalist movies are ideological agglomerations that draw on and perpetuate a range of contradictory discourses concerning the relationship between human beings and the environment.

Green Screen is divided into three parts, according to broad differences in environmental subject matter. The Introduction examines the way Hollywood cinema has tended to represent environmental issues according to the conventions of melodrama, and speculates on the implications for environmentalist politics of such aesthetic strategies. Part One explores the continuing symbolic role that wilderness plays in American popular cinema. Chapter One examines the way in which Hollywood movies have constructed 'nature' as a site of ecological concern, while perpetuating romantic desires for wilderness as a pristine, sacramental space. Chapter Two develops this notion further in an exploration of the aesthetics of landscape cinematography. Chapter Three analyses the gender implications of Hollywood's representation of non-human nature, while Chapter Four focuses on questions of ethnic difference, particularly as manifested in the recurring figure of the ecological American Indian. Chapter Five brings many of these themes together in a study of Hollywood movies set in the Amazonian rain forest.

Part Two of the book explores the representation of wild animals in Hollywood cinema, analysing the symbolic meanings projected onto them in American culture, and speculating on the implications for environmental politics of such anthropomorphic representational strategies. Chapter Six is concerned with the emergence of anti-hunting narratives in Hollywood cinema from the 1950s, while Chapters Seven and Eight trace the changing symbolic roles played by those 'stars' of modern conservationism—dolphins, orcas, wolves and bears—from monsters or varmints fit only for eradication, to idealized representatives of a benevolent wilderness that must be preserved. Chapter Nine examines Hollywood cinema's reconceptualization of Africa and its wild fauna in the light of modern conservationism.

Part Three deals with issues of development, land use and technology. Chapter Ten explores representations of agrarian and urban spaces in film, from the role of the family farmer to the ecological problems of urban environments. Chapter Eleven focuses on the ecological implications of automobile culture in Hollywood film, and Chapter Twelve focuses on movies which dramatize the hazards of nuclear energy.

Green Screen analyses these themes in Hollywood cinema by attempting to synthesize two approaches within film studies: close textual analysis and the general survey. The first critical strategy is useful for exploring the polysemic complexity of a small number of films. Nevertheless, it can have the drawback of being misleadingly narrow in its focus, in that it does not create a sense of how the chosen texts are representative of the full range of texts produced within a given culture. The second approach, which explores similarities and differences between a larger number of texts, can be more productive in this respect. Moreover, given the relative scarcity of critical work on cinema and environmentalism, *Green Screen* covers as large a field as possible in order to indicate directions for further research.

Inevitably, the selection of films covered in the book can be challenged. Films have been chosen for being either exemplary or typical, irrespective of the size of their budget or their box-office or critical success. Though the book mainly considers films as mediators of social issues irrespective of their artistic value, such value judgements inevitably inform all critical work, at least implicitly. The making and justifying of value judgements is, however, not the main intention of the book. Nor does it speculate at length on the 'influence' that the movies it explores may have had on attitudes to the environment in the United States or elsewhere.

Nor is the book intended as an adversarial polemic for a particular theory of environmentalism. Nevertheless, the approach taken in *Green Screen* doubtless favours certain critical stances more than others. In particular, its theoretical approach to non-human nature tends to endorse a critical realist position, as articulated for example by philosopher Kate Soper, and thereby seeks to distance itself from the more anti-empirical, anti-scientific and extreme social constructionist tendencies of some poststructuralist thinking. Soper points out that the epistemological confusion in extreme social constructionist ideas of nature lies 'in supposing that because we can only refer in discourse to an extra-discursive order of reality, discourse itself constructs that reality'.[5] She goes on to formulate the critical realist conception of nature that has been used as the basis for this book:

> the nature whose structures and processes are independent of human activity (in the sense that they are not a humanly created product) and whose forces and causal powers are the condition of and constraint upon any human practice or technological activity, however Promethean in ambition (whether, for example, it be genetic engineering, the creation of new energy sources, attempted manipulations of climatic conditions or gargantuan schemes to readjust to the effects of earlier ecological manipulations). This is the 'nature' to whose laws we are always subject, even as we harness it to human purposes, and whose processes we can neither escape nor destroy.[6]

It is this nature, independent of and external to human beings, that is the ground for the historically and culturally varied 'constructions' of its meanings that are the subject of this book. *Green Screen* seeks to identify the complex ways in which both non-human nature and the built environment have been conceptualized in American culture, and to analyse the interplay of environmental ideologies at work in Hollywood movies, while ultimately keeping the debate over environmental politics open and provisional.

Introduction

Melodrama, Realism and Environmental Crisis

Greenpeace video co-ordinator Karen Hirsch reportedly responded with horror at Sylvester Stallone's plan (subsequently abandoned) to make a movie in which Rambo takes on a band of environmental criminals. 'The issues are extremely complicated', she commented, 'they're not *supposed* to be black and white'.[1] A similar concern that the realities of environmental degradation are prone to be misrepresented by the conventional forms of the mass media is shown by social theorist Barbara Adam, who notes that the effects of environmental hazards such as ozone depletion, global warming, nuclear radiation and toxic pollution all tend to be slow to develop and are not amenable to simple or fast solutions, while their causes are invisible and systemic, and thereby complicate questions of individual and collective responsibility and liability.[2] In the face of such complexities, the aesthetic strategies of Hollywood cinema may indeed appear inadequate. Moreover, that Hollywood movies oversimplify complex social and political issues, and provide facile resolutions to real-life problems, has, of course, been a familiar complaint voiced by audiences throughout the history of American cinema. As Shohat and Stam argue, film spectators 'come equipped with a "sense of the real" rooted in their own experience, on the basis of which they can accept, question, or even subvert a film's representations'.[3] In recent years, interested groups have criticized environmentalist movies for their lack of correspondence to what they understand as the real world. Indeed, groups who feel they have been misrepresented by such movies have a history of public complaint, examples of which occur throughout this book: the nuclear industry challenged *The China Syndrome* (1978), for example, while the oil industry, Greenpeace and Native Alaskans all took exception to *On Deadly Ground* (1994).

These popular accusations of misrepresentation presuppose a realist interpretative context, in which a film is judged against a particular conception of reality, and is found wanting. Such demands for realism may find dissatisfaction in the tendency of Hollywood movies towards melodrama, defined by Linda Williams as a form 'that seeks dramatic revelation of moral and emotional truths through a dialectic of pathos and action'. Such aesthetic strategies are 'false to realism', and consequently open to accusations of misrepresentation.[4] With regard to environmental issues, negative criticism of melodrama, such as that of Greenpeace towards the prospect of a Rambo movie about environmental crime, takes two main forms. Firstly, the tendency of melodrama to construct environmental issues as individualized, Manichean conflicts between one-dimensional villains and heroes is seen to simplify the complex, often ambiguous allocation of blame and responsibility in such matters. Secondly, the closure effected at the end of a melodramatic fiction, when the hero resolves the narrative problem through decisive action, may appear too pat and glib a response to environmental crises which, in the real world outside the cinema, do not have their loose ends neatly tied up.

Yet Linda Williams places in a positive light the controversies that perennially surround Hollywood's melodramatic treatment of history and politics. Part of the 'excitement' of melodrama as a mode, she writes, 'is the genuine turmoil and timeliness of the issues it takes up and the popular debate it can generate when it dramatizes a new controversy or issue'. She defends the 'wish-fulfilling impulse towards the achievement of justice' in melodrama that 'gives American popular culture its strength and appeal as the powerless yet virtuous seek to return to the "innocence" of their origins'.[5] The next section will explore further the ways in which environmental conflicts are mediated in Hollywood cinema by the melodramatic mode.

The environmental politics of melodrama

Leo Braudy remarks that the heroes and heroines of what he calls the 'nature' movie of the 1980s were all figures identified with an authentic, primitive nature: ecologically sensitive tribal peoples (American Indians, aborigines, Neanderthals), children, women, animals, and psychics or natural wonders.[6] In explicitly environmentalist movies, these protagonists often enjoy a privileged, emotional and unmediated relationship with a re-enchanted nature. Moreover, heroic leadership in such films is often centred on a rebellious outsider, usually white and male, such as Forrest Taft (Steven Seagal) in *On Deadly Ground*, Jesse (Jason James Richter) in *Free Willy* (1993), or the maverick inventor Thomas Alden (Jeff Daniels) in *Fly Away Home* (1996), usually in alliance with family members or friends. In all of these movies, the trope of the reluctant outlaw-hero provides the means for representing ecological crises, in Robert Ray's terms, as 'short-lived' and 'solvable by decisive action', in keeping with familiar American mythological patterns.[7]

Environmental movies oppose the humanitarian innocence of their heroes to the commercialized values of their villains, for whom greed tends to be the prime motive. Environmental villainy takes two main forms. Firstly, hunters are often represented as the main obstacle to wild animal conservation. Once a heroic type in Hollywood cinema, the white hunter is now, with occasional exceptions, one of its arch villains. The second recurrent villain in the environ-mental movie is the representative of big business: the property developer, oil tycoon or nuclear plant manager. Environmentalist movies visualize the destructive effects on the environment of corporate capitalist greed in images of industrial technology as impersonal and unemotional. In particular, the noisy, brightly coloured bulldozer features as an impersonal and artificial destroyer of beautiful natural landscapes and traditional communities in conservationist movies as diverse as *The Milagro Beanfield War* (1988), *Fly Away Home*, *Meet the Applegates*, *FernGully* and *Medicine Man* (1992).

The framing of environmental issues as Manichean conflicts raises two important political questions: firstly, concerning the assumption of moral innocence on the part of the heroes, and secondly, the tendency of melodrama to individualize social conflicts. These questions will now be addressed in turn.

The melodramatic construction of heroism in terms of moral innocence and what Linda Williams calls 'virtuous suffering' has important implications for the question of responsibility and blame in environmental matters.[8] Sociologist Greg Myers, discussing the representation of political agency in the American children's environmentalist television show *Captain Planet*, argues that if 'villains are at the root of the evil, then environmental wrongdoing is removed from everyday actions'. [9] By making a melodramatic distinction between virtuous heroes and evil villains, *Captain Planet* allows for the complicity of its young audience in environmental degradation to be conveniently denied.

A related melodramatic strategy is the displacement of blame and responsibility for environmental degradation onto a generalized 'they' or 'we'. This rhetorical move also prevents the recognition of possible complicity in environmental problems, and also obfuscates the complex causality of those problems. In *The River Wild* (1994), Gail (Meryl Streep), a former river-guide dressed in a 'Save the Earth' T-shirt, wants to show her son the river 'before they ruin it'. This attribution of blame to a nameless and inaccessible 'they' is a consistent element in the depoliticization of environmental issues in Hollywood film. Gail's own possible complicity in environmental degradation, as a tourist using nature for recreational purposes, is conveniently denied. The obverse of this strategy is the framing of environmental issues as the responsibility of humanity as a whole. The end title of *Roar* (1981) informs the audience that 'we' are responsible for the decline in the lion population of Africa. That a homogeneous 'we' are all equally responsible is a form of mystification that also evades complex political issues of responsibility, liability and complicity by eliding social differences such as class, race, gender and geographical location. Given conditions of global inequality and exploitation, there can be, as Kate Soper puts it, no 'general species accountability' for environmental harm.[10]

The political problems raised by the construction of heroism and villainy in melodrama are matched, for Marxist critics Michael Ryan and Douglas Kellner, by the tendency for Hollywood movies to formulate social and political problems as conflicts between individuals, and thereby to endorse liberal solutions to those problems, preventing proper recognition of the need for more collective forms of political action. They argue, for example, that *The China Syndrome* is a failure in political terms, in that, by representing corporate power as individual rather than systemic, it 'fosters the rejection of the one solution to the social problems which liberalism so unsuccessfully addresses'.[11] Nevertheless, the writers concede that the movie's personalization of big business 'aids the enlistment of audience identification even as it misrepresents the reality'.[12]

There are, however, other ways of evaluating the way in which social and political conflicts are dramatized by the melodramatic mode. Richard Slotkin, for example, is more open to the aesthetic strengths of melodrama as a form that he views not simply as a failed or inadequate form of realism. In contrast to Ryan and Kellner, Slotkin argues that the construction of social conflicts in individual terms does *not* necessarily individualize power relations. Instead, he points out that the relationship between individual character types in a work of popular fiction can stand for complex, systemic power relationships. 'The hero's inner life—his or her code of values, moral or psychic ambivalence, mixtures of motive—reduces to personal motive the complex and contradictory mixture of ideological imperatives that shape a society's response to a crucial event', he writes. 'But complexity and contradiction are focused rather than merely elided in the symbolizing process.'[13]

Popular fictions, according to Slotkin's analysis, dramatize ideological contradictions and work out possible resolutions to them. This understanding of melodrama as a mediator of social contradictions is a useful way of discussing the representation of political issues in environmentalist movies. For example, the image of a group of chanting, placard-waving protesters features in *Fly Away Home*, *On Deadly Ground* and *Free Willy 2* (1995) as a signifier of collective dissent. Although these scenes are brief, and the narrative soon returns to the actions of the central protagonists, the suggestion of a link with larger, collective protests is nevertheless made. Accordingly, the politics of these environmentalist movies can thus be understood as reconciling individual, non-conformist protest with a broadly populist politics which endorses collective opposition to the destructive forces of corporate monopoly and elitist managerialism.[14] The melodramatic mode in this way provides a dramatic focus for ideological conflicts central to American society.

The following section will explore the complex interaction between melodrama and realism in two very different movies concerned with environmental crisis: *Day of the Animals* (1976), which takes ozone depletion as its starting point, and *A Civil Action* (1998), based on a real-life case of toxic contamination in Woburn, Massachusetts.

Environmental crisis in *A Civil Action* and *Day of the Animals*

The term 'realism' has, of course, been much debated within film studies, and is the counter-term to not only 'melodrama', but also 'fantasy' and 'modernism'.[15] The Hollywood movies discussed in this book may be thought of as variously located in a constellation between these key critical terms, the meanings of which are themselves not singular or fixed. In one region of the constellation lies *A Civil Action*, a movie which, although drawing mainly on a realist mode, also employs melodramatic conventions. At another end of the constellation is *Day of the Animals*, which draws on the genres of fantasy and horror to dramatize the effects of environmental catastrophe.

Realism in *A Civil Action* is signified in several ways. Firstly, the movie foregrounds its origin in an extra-filmic, real-life referent: lawyer Jan Schlichtmann's fight on behalf of eight families in Woburn, Massachusetts, whose children died of leukemia allegedly after drinking water contaminated with toxic waste dumped by two companies, W.R. Grace and Beatrice Foods. Indeed, Disney-owned Touchstone Pictures refused to change the names of the key protagonists in the case. Producer Rachel Pfeffer commented: 'If you change names, you have to start changing history. To be able to say this was based on a true story was important to the film-makers and the studio'.[16]

What the producer did not say, however, is that the movie is based on Jonathan Harr's account of the Woburn trial: it is a mediation of a mediation of history. Also ignored is the fact that Harr's account was itself criticized by other parties in the case; firstly for playing down what other commentators saw as Schlichtmann's mishandling of the case, and secondly, for concentrating on the lawyer's side of the story, rather than that of the victims.[17] The 'reality-effect' of *A Civil Action*, then, is the product of selection and subjective evaluation of historical information, as well as aesthetic decisions over both narrative 'alignment' (the story is told from Jan Schlichtmann's point-of-view) and audience 'allegiance' (the audience comes to sympathize and 'root for' Schlichtmann, as played by John Travolta).[18]

The second way in which *A Civil Action* signifies itself as realist is through audio-visual and narrative conventions. As Noël Carroll observes, 'realism' in cinema implies not a direct, unmediated correspondence between a representation and reality, but a set of stylistic choices. Realism, he writes, 'is not a simple relation between films and the world but a relation of contrast between films that is interpreted in virtue of analogies to aspects of reality'. There are, then, several types of cinematic realism, including Soviet realism, deep-focus realism and Neorealism, none of which 'strictly correspond to or duplicate reality, but rather make pertinent (by analogy) aspects of reality absent from other styles'.[19] The stylistic choices that most often signify realism include, as Bordwell and Thompson note, 'authenticity in costume and setting, "naturalistic" acting, and unstylized lighting'.[20] In *A Civil Action*, then, realism is signified through deep-focus cinematography, location shooting, under-stated 'naturalistic' acting, and a screenplay that prefers complex but low-key action to sensationalism, and graduated characterization to Manichean caricature.[21]

Nevertheless, elements of melodrama are central to the movie's dramatization of Harr's book. The screenplay adds a simple arc of character development to the central protagonist which is absent from the book, as Schlichtmann begins the film arrogant and selfish, but is transformed into a man of conscience, learning through his experiences that, in the words of the magistrate at his bankruptcy hearing at the end of the film, 'the things by which one measure's one's life' are ultimately more than financial and materialistic. This familiar narrative formula combines with the charisma of Travolta's star performance to turn the movie towards melodrama in spite of its rhetoric of realism. Moreover, after the settlement with the food company Beatrice, the movie invents a scene in which Schlichtmann parts company with his firm and 'goes it alone'. 'You always went your separate way, Jan', he is told by his colleague, thereby becoming another familiar type in Hollywood melodrama: the individualistic, heroic outsider battling an impersonal bureaucracy. When Schlichtmann's perseverance finally unearths evidence of a cover-up at the Woburn site, he summarizes in a voice-over what he has learned from his experiences: 'if you calculate success and failure, as I always have, in dollars and cents divided neatly into human suffering, the arithmetic says I failed completely. What it doesn't say, is if I could somehow go back, knowing what I know now, knowing where I'd end up if I got involved with these people, knowing all the numbers, all the odds, all the angles, I'd do it again.' These words are accompanied by ethereal choral music on the soundtrack, suggesting thereby the apotheosis of the lawyer-turned-hero: Schlichtmann, as played by Travolta, a flawed but charming man redeemed and made noble by self-sacrifice. In Hollywood cinema, Linda Williams notes:

> supposedly realistic cinematic *effects*—whether of setting, action, acting or narrative motivation—most often operate in the service of melodramatic *affects*. . . . If emotional and moral registers are sounded,

> if a work invites us to feel sympathy for the virtues of beset victims, if the narrative trajectory is ultimately more concerned with a retrieval and staging of innocence than with the psychological causes of motives and action, then the operative mode is melodrama.[22]

If the representation of environmental crisis in *A Civil Action* is achieved through a combination of both realist and melodramatic techniques, then the science-fiction-horror movie *Day of the Animals* (1976) lies at the other end of the constellation between realism and melodrama, and realism and fantasy. The plot of the movie cleverly exceeds everyday notions of realistic plausibility: increasing levels of solar radiation caused by the hole in the ozone layer have triggered a virus which causes wild animals to attack a group of backpackers in the High Sierras. Nevertheless, the movie conforms to the conventions of science fiction by placing this narrative within a rhetoric of 'scientific' plausibility, established by the opening title:

> In June 1974, Drs. Frank Sherwood and Mario Molina of the University of California startled the scientific world with their finding that fluorocarbon gases used in aerosol spray cans are seriously damaging the Earth's protective ozone layer.
>
> Thus, potentially dangerous amounts of ultra-violet rays are reaching the surface of our planet, adversely affecting all living things.
>
> This motion picture dramatizes what COULD happen in the near future, IF we continue to do nothing to stop this damage to Nature's protective shield, for life on this planet.

Writing of the science fiction movie *The Thing* (1982), Steve Neale observes that the film is 'involved both in establishing its own credibility, and in establishing its own regime of credence—the rules, the norms and the laws by which its events and agents can be understood and adjudged. What is probable or possible in this world? How does it operate? What is regarded within it as unusual, unlikely, inexplicable?'[23] Different genres and fictional modes, then, rely on different types of motivation and justification for their fictional events. The rhetoric of realistic plausibility employed at the start of *Day of the Animals* is thus typical of a science fiction 'what if' scenario, in the way it draws on hyperbole (the ultraviolet rays are 'adversely affecting all living things') and an obfuscation of the difference between possibility and probability ('what COULD happen in the near future'). Clearly, if such playful strategies are judged according to the criteria of plausibility normally applied to a more realist film such as *A Civil Action*, the movie may be judged inadequate to its initial premise about ozone depletion.

Indeed, such a negative evaluation of the film was made in the anonymous review in *Variety*, which noted that the appearance of the opening card merely provoked laughter in the audience. 'Hitchcock, by the way', the reviewer continued, 'never stooped to explain why his feathered characters went wild in "The Birds", which was one reason why his pic was so genuinely scary. Once it's blamed on spray cans, it all seems mundane and silly'.[24] What is 'silly' in the film, and recognized by the audience's ironic laughter, is the incongruity of scale between the cause given by the movie (spray cans) and its effect (mad animals on the rampage). Comedy arises as the audience recognizes that the film fails to motivate in a satisfactory way, even within what Neale calls its own 'regime of credence', the incongruity between the phenomenon of ozone depletion, whose effects as understood by science are slow, subtle and spatially dispersed, and its fictional representation in the narrative as simple, localized and fast-acting. 'I told you that sun seemed damned peculiar today', says the policemen in the film, when radio reports about the hole in the ozone layer begin to come in to the town in the High Sierras from all over the world.

Despite the hostile critical reception of *Day of the Animals*, it is nevertheless important to recognize the artistic strengths of melodrama and fantasy as ways of mediating environmental issues. As Steve Neale notes, melodramas can have a political power that transcends the more culturally respectable forms of naturalism and verisimilitude.[25] A more positive interpretation of *Day of the Animals*, then, may view it as a symbolic narrative which visualizes the potentially disastrous effects of a process (ozone depletion) that is invisible and abstract. In doing so, the film repeats a recurrent motif in horror and disaster movies: that of the revenge of nature on the human beings that have harmed it.[26] Maurice Yacomar notes that stories of 'natural attack' in the disaster genre dramatize 'people's helplessness against the forces of nature', while animal attack films 'provide a frightening reversal of the chain of being, attributing will, mind, and collective power to creatures usually considered to be safely without these qualities'.[27] *Day of the Animals*, accordingly, dramatizes the monstrous return of nature as the repressed of modern industrial society.

The apocalyptic character of such a narrative is, however, politically controversial for some commentators on environmentalism. Socialist environmental writer Tom Athanasiou argues that the use of apocalyptic rhetoric in non-fictional books such as Paul Ehrlich's *The Population Bomb* (1968) and Bill McKibben's *The End of Nature* (1989) is ultimately a symptom of political despair rather than of radical empowerment, and makes environmentalists easy targets for attacks from right-wing anti-environmentalists.[28] In contrast, M. Jimmie Killingsworth and Jacqueline S. Palmer defend the use of apocalyptic rhetoric in non-fiction texts, including Rachel Carson's *Silent Spring* (1962), for what they see as its subversive political function. Apocalyptic texts, they argue, by presenting 'worst-case scenarios' as 'foregone conclusions', constitute a radical attack on the notions of progress held by big business, big government and big science. The 'image of total ruin and destruction', they write, 'implies the need for an ideological shift'.[29] While extending this argument to a discussion of Hollywood cinema is initially problematic, in that fictional movies clearly do not make claims regarding empirical truth in the same way as the non-fictional texts cited above, the notion that apocalyptic fictions can also articulate a desire for radical social change, or at least a protest against the status quo of big business and big science, is nevertheless worth exploring.

A recurrent tendency that may be discerned in such fictions is that the 'image of total ruin and destruction' to which Killingworth and Palmer refer is usually averted at the end of the film. Thus the dystopian science fiction film *Soylent Green* (1973) ends on a freeze-frame of Charlton Heston's upraised finger, as he pleads for further action to restore civilization from the grip of corporate capitalism, which has degenerated into cannibalism. This combination of affirming the power of the individual to change the future, with a relatively open and unresolved sense of that future, is a typical strategy for movies dealing with environmental apocalypse, as will be seen in discussions of *The China Syndrome* and *On Deadly Ground*, for example, later in this book.

It is in comparison with such relatively open endings that the resolution of *Day of the Animals* appears problematic. As the corpses pile up, and law and order breaks down, the Environmental Protection Agency suddenly announces on the radio that the ozone level 'continues to correct itself', and that the 'virus mutation infecting human and animal life is unable to sustain itself as the sun's radiation decreases to normal levels'. The radio announcement concludes: 'All altitudes will be completely safe within twenty-four hours.' Although this reversion to a state of equilibrium conforms to the normative expectations of the horror genre, the abruptness with which nature heals itself seems evasive given the literally global extent of the problem posited at the start of the film. Moreover, environmental catastrophe has been averted contingently, rather than through the action of the characters in the story. It is the abrupt and arbitrary nature of the movie's closure, then, that appears pat, trivializing and exploitative of the seriousness of its environmentalist premises.

Such incongruous endings are, of course, common to Hollywood movies, which, as Maltby and Craven write, are marked by 'the dynamic reciprocity between the sometimes pat resolutions of individual stories and the frequently gaping irresolution of their social implications'.[30] Ultimately, then, *Day of the Animals* displaces and contains the apocalyptic anxieties it raises. In this sense, the 'need for an ideological shift' spoken of by Killingsworth and Palmer as a positive implication of apocalyptic texts is itself dissipated. Entertainment, observes Michael Wood, is not 'a full-scale flight from our problems, not a means of forgetting them completely, but rather a rearrangement of our problems into shapes which tame them, which disperses them to the margins of our attention'.[31] In its displacement of the environmental issues with which it began, *Day of the Animals* is typical of many of the environmentalist movies discussed in this book. Significantly, the initial premise of ozone depletion is used as a basis for the more familiar thematic concern, that of individual and group survival under competing forms of male leadership, with which the rest of the narrative is mainly concerned. As the following chapters demonstrate, Hollywood environmentalist movies often use their concerns with non-human nature, whether wilderness or wild animals, as a basis for speculations on human social relationships, thereby making those concerns conform to Hollywood's commercial interest in anthropocentric, human interest stories.

The rest of this book will seek to place the films under discussion within the constellation of realism, melodrama and fantasy outlined in this Introduction. The important questions that will be asked of such films, whether a more realist film such as *A Civil Action*, or a more melodramatic one such as *Day of the Animals*, will concern, in the words of Stephen Prince, 'the kinds of linkages that connect the represented fictionalized reality of a given film to the visual and social coordinates of our own three-dimensional world,' an inquiry, he notes, that 'can be done for both "realist" and "fantasy" films alike'.[32]

I

WILDERNESS IN HOLLYWOOD CINEMA

ONE

Discourses of Nature and Environmentalism

The Hollywood movies examined in this book draw on and combine a range of different environmentalist discourses, from conservationism to preservationism, and mainstream to radical environmentalism. It is useful at this point to offer a brief overview of these discourses, before undertaking an analysis of how they are mediated by the films themselves.

Conservationism, since its origins in Progressivism at the turn of the nineteenth century, has taken a utilitarian attitude to non-human nature, treating it as a resource to be managed and developed for use and economic profit. In contrast, preservationism has argued for the need to preserve wilderness as a realm of spiritual and aesthetic contemplation separate from resource use.[1] With the rise of modern environmentalism in the early 1960s, conservationism has become the 'mainstream' ('reform', 'moderate', or 'shallow') wing of environmentalism. Mainstream environmentalism continues to place environmental concerns within the needs of a capitalist economy to sustain commodity consumption, profit maximization and economic growth, by calling on the expert knowledge of economists, engineers and scientists to provide *ad hoc*, technical solutions to environmental problems. For example, the addition of catalytic converters to automobile exhausts is a key mainstream environmentalist proposal to address the problem of air pollution. In being defined as technical rather than political, environmental problems are viewed as solvable within the existing system of capitalist bureaucratic-technocratic rationality administered by the state and private corporations. Advocates of mainstream environmentalism argue that these solutions are practical, pragmatic and realistic, and are therefore the most effective form of environmental restoration.

Radical environmentalism includes a range of different approaches, from deep ecology to social ecology and ecofeminism. The broad area of agreement between these groups is that mainstream environmentalism is ultimately counter-productive, in that its attempts to strengthen capitalism simply perpetuate one of the fundamental causes of ecological decline itself. According to radical environmentalists, mainstream environmentalist faith in the reform potential of technology is also misguided. Moreover, they argue that by depoliticizing environmental issues, mainstream environmentalism prevents the emergence of more radical or revolutionary environmental politics based on notions of social justice.[2] Mainstream environmentalism relies instead on 'greenwashing', or the attempt to deny or cover up the fundamental causes of environmental degradation. Socialist environmentalist Tom Athanasiou defines 'greenwashing' as a mainstream strategy in which 'images of change substitute for and exaggerate change itself'.[3]

Marxist David Harvey draws on Herbert Marcuse's notion of 'repressive tolerance' to argue that mainstream environmentalism is in the process of incorporating more radical and oppositional environmental ideologies for its own benefit. What he calls a 'limited articulation of difference' in official environmental discourses thus plays a 'sustaining role for hegemonic and centralized control of the key institutional and material practices that really matter for the perpetuation of capitalist social and power relations'.[4] Harvey contends that prospects for environmental restoration and social justice are set back by the incorporation of radical ecology into mainstream environmentalism, because the latter is thereby strengthened. In contrast, anthropologist Martin W. Lewis argues that the incorporation of radical ecological thinking by mainstream environmentalism is bad for the prospects of environmental restoration, not because it *strengthens* mainstream environmentalism, but because it weakens it. For Lewis, radical environmentalism is itself counter-productive, particularly in what he sees as its anti-scientific, romantic and technophobic tendencies.[5]

The main intention of this book is not so much to adjudicate between these contending theories, as to analyse the ways in which particular Hollywood movies mediate such ideologies in often complex, contradictory and incoherent ways. The rest of this chapter will therefore examine the different constructions of non-human nature in Hollywood cinema, and speculate on the implications they hold for environmentalist politics.

Conservationism and the western: *Valley of the Giants*

Although the popularity of the western genre coincided with the emergence of federal conservationism in the early years of the twentieth century, few westerns developed an explicitly conservationist stance towards contemporary struggles over land use. Instead, the classical western celebrated Manifest Destiny, and the settlement and development of American land for ranching and agriculture. The wilderness was turned into a garden through the heroic work of the pioneers, supported by the justified violence of the white male hero. Nevertheless, the western was often ambivalent in its attitude to the transformation and development of what it took to be pristine wilderness. John Ford's *The Iron Horse* (1924) celebrated the epic unification of the American nation by the transcontinental railroad. Yet the same director's *Stagecoach* (1939) displayed for the Depression era a nostalgic yearning for the unspoiled wilderness, and romanticized its avatar, the Ringo Kid (John Wayne), as free of the corruptions associated with small-town 'civilization'. Despite such ambivalence, however, the classical western, as Robert Ray shows, drew on the myth of American exceptionalism to assume a future for the nation that was open and without limits.[6]

An early exception to the omission of conservationist issues from the western genre was a series of movies based on Peter B. Kyne's novel *The Valley of the Giants* (1918). These movies attempted to reconcile a desire to preserve wilderness as a space for spiritual contemplation with its use as a resource for capitalist expansionism. The first film version of *The Valley of the Giants* was made in 1919. Kyne's book was subsequently filmed three more times, in 1927, 1938 and 1952, the latter version as *The Big Trees*.[7]

The second version of *Valley of the Giants* (1927) begins with a title that alludes to the nationalistic cult surrounding the giant redwood tree in American culture: 'A century after the Declaration of Independence, the giant redwoods of California were still a forest primeval—virgin, venerable and awe-inspiring'. As historian Simon Schama has shown, visitors from the East influenced by Transcendentalism saw the Sierra Nevada redwood (*Sequoia-dendron gigantea*), or Big Tree, as a visible sign of the presence of God in nature, and of divine sanction for American national interests. The gigantic size of the Big Trees proclaimed, in Schama's words, 'a manifest destiny that had been primordially planted; something which altogether dwarfed the timetables of conventional European and even classical history'.[8] The Big Trees thus became an early cause for nature preservationists wishing to preserve them for the access they granted to moral and spiritual enlightenment. These calls for preservation were helped by the fact that, unlike the coastal redwood (*Sequoia sempervirens*) to which early botanists mistakenly related it, the Sierra Nevada redwood is too brittle for timber.

In any case, as Nancy K. Anderson observes, when 'the supply of big trees was perceived as unlimited, all uses were sanctioned'.[9] As its second title made clear, then, the 1927 version of *Valley of the Giants* was able to evoke the symbolic connotations of the Big Trees while at the same time celebrating the clearing of the Californian forests by a heroic American pioneer:

> In the Valley of the Giants, John Cardigan built his home—amid trees that were old when Ancient Rome was new. Through fifty years he toiled—heaving an empire from the wilderness—building a city, with lumber mills employing thousands.

John Cardigan (George Fawcett) has preserved the Valley of the Giants, 'the finest redwoods in the world', as a memorial to his late wife. As the camera pans upwards from her grave past the trees to the sky, it constructs a cinematic version of the image of the redwood forest as a natural cathedral, with divine sunlight shining through the trees, popularized in the nineteenth-century photographs of Carelton Watkins and the paintings of Albert Bierstadt.[10]

When John Cardigan's son Bryce (Milton Sills) arrives from New York to inherit his father's lumber business, the movie further reconciles the desire to preserve the Valley of the Giants with the successful running of the logging company. Bryce's greedy business rival Pennington (Charles Sellon) has designs on the Valley of the Giants, and therefore refuses to renew the hauling contract that allows Bryce to use his railroad to export his logs. However, the soft, Eastern greenhorn toughens up sufficiently to survive and prosper in the West. Furthermore, although Pennington accuses the Cardigans of spoiling their workers, the movie celebrates Bryce as a paternalistic capitalist, who is able to draw upon the loyalty of his workers, as well as their self-interest in defending their jobs and homes, to build his own railroad, and thereby continue the Cardigan family business. As Bryce says of his workers: 'I love these men and their families as if they were my own children'. In addition to his benevolent treatment of his workers, it is the sentimental attitude that Bryce shows towards his mother's redwood memorial that guarantees that the form of capitalism he practices is beneficent. In terms of contemporary environ-mentalism, the movie's reconciliation of redwood preservation with capitalist growth is an endorsement of Gifford Pinchot's official conservationist policy of managing forests for use through selective cutting, which he initiated when he assumed leadership of the Division of Forestry in 1898.[11]

Appropriate to its New Deal context, the 1938 version of *Valley of the Giants*, remade for sound by Warners, was more sceptical than its predecessor of the virtues of unregulated capitalism. The movie opens with a title that also states a more explicit and urgent conservationist agenda than the 1927 movie:

> The Pacific slopes of California—where, like living cathedrals, giant redwoods stand as an heritage of beauty symbolizing in their grandeur man's hope for immortality—but man in his greed saw in these trees only a source of profit and soon—backed by limitless power and wealth the timber barons moved in, ruthlessly crushing small land owners, destroying human happiness and beauty alike. It was the era of the timber steal.

In this version of the story, Bill Cardigan (Wayne Morris) has vowed to conserve the Valley of the Giants as a memorial for his late parents, who are buried there. He wants to preserve the redwood park as a place for contemplation in a materialistic society. 'People change when they come in contact with the woods', he says. 'The trees make 'em stop and think about a lot of things they've been too busy to notice before'.

In contrast to Cardigan's reverence for the Big Trees, Howard Fallon (Charles Bickford) is a land grabber and claim jumper, whose attitude is unsentimental and mercantile. 'Trees are trees', he says. 'If you've seen one, you've seen them all'. The greedy and exploitative Fallon tells Cardigan of his ambition to fell more trees in order to make more money. More explicitly than in the 1927 movie, Cardigan's response reflects official federal conservationist policies that called for the planned management of trees as a sustainable natural resource:

> Cardigan: The way you operate, these forests would be cleaned out in ten years.
> Fallon: What of it? There's a lot of other places to move on to.
> Cardigan: We don't figure to move on. There'll always be enough timber here, with reasonable cutting and planned reforestation.
> Fallon: Reforestation? What do you care what's growing fifty years from now?
> Cardigan: You wouldn't understand that, Fallon. My father spent his whole life planning a safe future for this district and the friends who settled here with him.

The subsequent conflict between Cardigan and Fallon is resolved through a liberal interpretation of Cardigan's notion of 'reasonable cutting'. Fallon files a claim under the Timber Act in order to evict the homesteaders who are in the way of his logging scheme. He then buys Cardigan's note with the bank, and demands payment within six weeks. In response, Cardigan decides to increase the logging of his other trees in order to raise money to pay off the note, and thereby preserve the Valley of the Giants from Fallon's claim.

'Every time you sink your axe into a redwood', Cardigan encourages his workers, 'you're whittling Fallon down to our size'. The climax of the movie thus becomes a celebration of the heroic work of the logger, with scenes of logging accompanied by uplifting music on the soundtrack. However excessive and complacent this resolution appears in retrospect, when judged from the perspective of decades of deforestation in the American West, its enthusiastic endorsement of a notion of sustainable logging is in keeping with the official utilitarian conservationism of its day.

The most recent version of *The Valley of the Giants*, renamed *The Big Trees* (1952), shifted the focus of preservationist concerns from a logging family to a religious colony, which fights to save its giant sequoias from an illegal timber claim made by a Wisconsin syndicate led by Jim Fallon (Kirk Douglas). The sequoias are again viewed as both sacred and a source of national pride, and are signified as sublime by upward panning shots set to Christian choral music. In contrast to this reverential attitude to nature, Fallon is again greedy and materialistic, viewing the trees solely as commodities, a mind-set symbolized by his habit of measuring them with a ruler. The plot hinges on the decision of the religious colony to cut down and sell its smaller trees in order to pay the filing fees to claim their land back from Fallon, and thereby preserve their sacred redwoods. As in the previous versions of the story, therefore, the conflict over redwood preservation is resolved in a way that celebrates economic growth and development as forces of social good.

Despite this endorsement of the commodification of the Big Trees, however, the figure of the logger in all of the movies based on Kyne's *The Valley of the Giants* remains ambiguous: logging is seen as a noble profession when it upholds official conservationist policies and sentimental attitudes to nature, but not when practiced by unregulated, unscrupulous businessmen. The main difference between the movies and official conservationist polices, however, is that none of them makes a case for either the public ownership or federal regulation of forests. Instead, in keeping with the conventions of the western genre, they valorize small, private ownership of land and the need for individual, vigilante action to protect it. Their hero is therefore the good outlaw who reluctantly resorts to justified violence in order to protect innocent people against robber barons who manipulate the law for their own corrupt vested interests. The environmental politics of this cycle of movies is therefore ultimately shaped by the melodramatic conventions of the western genre.

Disney's *Bambi* and the 'balance of nature'

Scientific journalist Stephen Budiansky examines three questionable assumptions about non-human nature that have become commonplace in both radical and mainstream environmentalist thought, and are also perpetuated in the artefacts of popular culture. 'Today', he writes, 'many nature lovers innocently believe that the "balance of nature", or the notion that every species is interconnected in a delicate "web of life" that will collapse if but a single strand is cut, or the idea that "nature knows best" how to manage itself are scientific statements of fact derived from modern ecological research'. These dubious notions, he argues, lead to the misleading conclusion that nature, 'if only it is left alone and freed from human interference, tends toward a state of harmony, balance, and beauty—and conversely, that wherever man treads is trouble'.[12]

The concept of the 'balance of nature', implying that nature is static, timeless and harmonious, has recently been challenged not only by scientific ecologists but also by environmental historians. Both argue that natural systems should be more accurately thought of as dynamic phenomena which are more often than not the product of an ongoing, complex history of interactions with human cultures. The natural landscape is, in part at least, a human artefact, the historical product of millions of years of human transformations, such as forest clearances, agriculture, deliberate burning, and the importation of exotic plants and animals. Even old-growth forests are not simply 'natural' or 'virgin', but are the historical products of annual burning by indigenous peoples.

Moreover, far from balanced and harmonious, nature as understood by contemporary scientific ecology is chaotic and unstable.[13] Budiansky argues that the myth of nature as static, balanced and pristine has had a debilitating effect on the prospects for ecological restoration. 'Man's long history as an agent of change in nature and nature's own perverse tendency toward disorder and complexity', he writes, 'pose a complication that simple policies built upon the idea of nature's innate balance cannot even begin to cope with'.[14]

Environmental historian William Cronon also argues that sustaining biodiversity requires an active, interventionist attitude towards the natural environment, and is therefore hampered by romantic notions of wilderness as existing outside of human culture, and by the preservationist 'hands-off' arguments to which such notions invariably lead. 'To the extent that biological diversity (indeed, even wilderness itself) is likely to survive in the future only by the most vigilant and self-conscious management of the ecosystems that sustain it', he writes, 'the ideology of wilderness is potentially in direct conflict with the very thing it encourages us to protect'.[15]

In Hollywood cinema, Disney's *Bambi* (1942) stands out as a powerful construction of nature as pristine and harmoniously balanced. Moreover, as perceptive commentaries by Matt Cartmill and Ralph Lutts point out, the film essentializes all human activity as a violation of that natural harmony. Human beings are merely interlopers in nature, their presence in the movie reduced to the feet of the hunters who kill Bambi's mother and later start the devastating forest fire. Human beings bring only death and destruction to the pristine Eden inhabited by the benign and gentle wild animals.[16]

However, if the positing of human beings as unnatural violators of nature draws on preservationist attitudes, *Bambi* also reflects the conservationist ideologies of its day, particularly in its representation of the forest fire as simply destructive and undesirable. Official forestry policy at the time of the film's production advocated the prevention and fighting of forest fires, because, according to utilitarian conservationism, fires were unnatural and a waste of valuable resources.[17] In 1944, the Wartime Advertising Council used the image of Bambi in its fire prevention campaign, thereby appropriating the movie for its conservationist agenda. (Problems with licensing, however, led to Bambi's replacement the following year by the more familiar symbol of Smokey the Bear)[18]

However, in representing the fire as unequivocally destructive and unnatural, *Bambi* runs counter to modern ecological understandings of the role of fire within forest ecosystems. In the 1960s, ecologists began to understand the ecological usefulness of fire in renewing forest habitats. Regular burning releases minerals into the soil, and clears out old or diseased timber, thereby encouraging new growth and diversity. Some nuts and seeds actually need fire to sprout. Nowadays, as Patricia Limerick points out, the ideology of 'nature knows best' informs policies of fire prevention and control.[19] Forester Paul Schullery thus criticizes *Bambi* for perpetuating outdated ideas on the role of fire in nature, and asserts that the public reaction of horror to the National Park Service's policy of letting the 1988 fire in Yellowstone National Park burn itself out had 'a great deal' to do with *Bambi*. Schullery cites Roderick Nash's comment that *Bambi* did 'more to shape American attitudes towards fire in wilderness ecosystems than all the scientific papers ever published on the subject'. With the release of the movie on home video in 1990, Schullery feared that children will continue to learn ecological lessons 'that were discarded by fire ecologists decades ago, and those lessons are not good enough in today's environmentally attuned world'.[20]

In environmentalist terms, then, *Bambi* combines both preservationist and conservationist attitudes to the natural world. As an animal fable, moreover, it deploys its concept of natural order to naturalize a conservative view of human social relations, particularly in its patriarchal attitude towards gender relations.[21] As the next section will show, the same company's *The Lion King* (1994) does similar ideological work for a contemporary era in which a rhetoric of both scientific and deep ecology can give cultural authority to the movie's ideological construction of the natural order. The movie is a contradictory agglomeration of several constructions of nature, which serve both to promote a preservationist attitude to non-human nature, and to naturalize a conservative model of the human social and political world.

The Circle of Life: Disney's *The Lion King*

It is useful to place *The Lion King* within the context of Hollywood environmental advocacy at the end of the 1980s, when two much-publicized environmental advocacy organizations, the Earth Communications Office and the Environmental Media Association (EMA), were formed in Hollywood. The EMA was founded by a group of Hollywood producers, directors, actors and agents, including television producer Norman Lear and Disney chief executive Michael Eisner, and has been supported by actors such as

Christopher Reeve, Lindsay Wagner and Billy Crystal.[22] The organization's stated purpose is 'to mobilize the entertainment industry in a global effort to educate people about environmental problems and inspire them to act on those problems now'. This mission statement has two main applications: to promote 'greener' practices in Hollywood film production, and a greater recognition of environmental issues in film and television productions themselves. As far as the first area is concerned, the EMA promotes recycling and waste management practices in the Hollywood studios, lobbying, for example, for the recycling of lumber and the use of wood substitutes in set construction. The EMA also credits itself with an increase in references to environmental issues in films and television shows, both as incidental dialogue and as whole storylines. The organization provides an environmental research and fact-checking facility, as well as consultation on story ideas.[23]

Coinciding with Michael Eisner's time on the board of the Environmental Media Association at the start of the 1990s, the Walt Disney Company began working hard to gain a reputation for enlightened environmental policies, especially in its theme parks, which introduced policies of water conservation, solar energy, recycling, and organic pest management. Kate Moss Warner, director of horticulture and 'environmental initiatives' at Walt Disney World in Florida, was quoted as saying: 'From a moral and ethical standpoint—from Walt Disney himself—we have always had an environmental mission. . . . Now we want others to share in that mission'.[24]

However, the company's self-congratulatory 'sense of mission' has come under constant challenge from environmental and labour groups. Plans for the proposed theme park Disney America to be built in Virginia were postponed in 1994 after local environmental opposition.[25] The Disney Company has also come under pressure from environmental and labour activists over its business practices. The National Labor Committee's 'Open Letter to Walt Disney' accused the company of labour abuses in its licensed garment manufacturing factories in Haiti. Andrew Ross summarizes the basic economic inequalities: Michael Eisner earned 'over $200 million from salary and stock options in 1996, which, at $97,600 per hour, amounted to 325,000 times the hourly wage of the Haitian workers who made Pocahontas, Lion King, and Hunchback of Notre Dame T-shirts and pajamas, and who sewed on Mickey Mouse's ears'.[26]

The contradiction between Disney's environmentalist rhetoric and its actual performance thus suggests that the company is engaged in 'greenwashing' its public image, in exactly the way described by Tom Athanasiou at the start of this chapter. In this light, *The Lion King* itself may be seen as a 'greenwashing' movie, that incorporates ideologies from radical environmentalism, only to travesty and depoliticize them, and thereby perpetuate the interests of mainstream environmentalism at its most shallow.

The film's prologue sequence, 'The Circle of Life', uses the circle as a metaphor for a natural world in which all life is interconnected in a single, balanced, harmonious structure. The images that accompany the song connote inter-species co-operation: an elephant carries birds in its tusks and gazelles leap playfully, as the spectator's gaze swoops upwards following the hornbill Zazu in an exuberant celebration of natural harmony and inclusiveness. A close-up shows an ant carrying a pile of leaves, before zooming out to a wider shot of zebras running across the plains: the movement in the *mise-en-scène* enacting the interconnectedness between 'great and small' referred to in the song lyrics at that moment.

The image of the Circle of Life recalls Disney's True-Life Adventure films of the 1950s, which, as Alexander Wilson pointed out, were structured by natural cycles, always beginning with spring, sunrise and the birth of an animal. In *The Lion King*, this Judaeo-Christian iconography of cyclical time, recalling *Ecclesiastes* 3:1–8, is overlaid with an eclectic range of other discourses on nature, including the Plains Indian myth of the Sacred Hoop, the medieval Great Chain of Being, eighteenth century concepts of a divinely balanced nature, the 'food chain' of modern Darwinian ecological science, and deep ecology.[27]

As the animals gather to bow down to the new-born lion cub, the lyrics talk of the need to 'find our place' in the Circle of Life. In this way, the movie constructs an ecological rhetoric of nature in order to naturalize specific human power relationships: the Circle of Life conflates a Darwinian sense of 'place', that is, being part of what ecologists after Darwin call a 'niche', with *knowing one's place* in a social hierarchy based on conservative notions of power and authority.[28] The subsequent narrative confirms this confusion of social and ecological hierarchies. When the divinely ordained hierarchy presided over by the lion king Mufasa is disrupted by his brother Ska's usurpation of power, ecological disaster is unleashed on the land. Ska grants freedom to the hyenas, who, by rebelling against their 'place' in the social system, disrupt the food chain, and thereby destroy the environment and bring famine to the land. The movie in this way thus constructs a narrative that represents animals as part of an ecological system, and at the same time employs them as allegories in a human social and political fable. The hyenas are thus demonized because they are scavengers, rather than noble hunters like the lions. Moreover, not only do they violate the Protestant work ethic, but they are also identified vocally with proletarian African-Americans and Hispanics. Indeed, Terry Diggs notes that, as a political allegory, *The Lion King* conforms to contemporary Republican explanations of America's decline: 'Pride Rock's infrastructure goes down for the count when an effeminate pretender grabs the king's throne and turns the country over to handout-accepting ne'er-do-wells'.[29]

What is missing from the 'Circle of Life' prologue is central to the rest of the film: the existence of predation in nature. Mufasa later explains predation to his son Simba in terms of the 'food chain', in which all life, as he puts it, 'exists together in a delicate balance'. When he is king, Simba will need to 'understand that balance, and respect all the creatures, from the crawling ant to the leaping antelope'. When Simba questions this, his father explains that, although the antelope is prey to the lion, their relationship is cyclical and mutual: 'When we die our bodies become the grass, and the antelope eat the grass, and so we are all connected, in the great circle of life'.

The metaphor of the 'food chain' used here to suggest ecological interdependency has its origins in the work of British zoologist Charles Elton. As Donald Worster shows, the concept of the food chain differed from the earlier notion of the Great Chain of Being, which 'ranked all species on a single grand staircase, those at the top of the stairs being the most noble and honored'. In a 'food chain', in contrast, 'the bottom of the chain, rather than the top, is the most important link: the plants make the whole system possible'.[30] *The Lion King* conflates these two models of nature, transforming the amoral nature of Darwinian ecology into a moral nature that can thereby serve as a prescriptive model for the movie's conservative social and political agenda.

Critic Ted Kerasote criticizes *The Lion King* for finding no place for human beings in its Circle of Life. Whereas *Bambi* reduced human presence in nature to the hunters' feet, *The Lion King* excludes human beings completely from its self-regardingly 'green' vision of nature. 'In its attempt to discard the traditional imperialist role we have played in nature', writes Kerasote, '*The Lion King* creates a revisionist view that is nearly as dysfunctional and alienating'. The film, he continues, 'takes the increasingly fashionable preservationist attitude of seeing humans as passive ecotourists watching an Eden in which we play no part. In both cases, first as conquerors and now as observers, we are led away from the complex and often soul-searching involvement that being a true participant in nature's community entails'.[31] Nature exists in the film as a national park from which the Masai, the human inhabitants of the real Serengeti, are completely written out.

In this light, the deep ecological rhetoric of organic inter-connectedness which the movie evokes is particularly misleading. Anthropologist Stan Steiner described the Native American philosophy of the Circle of Life as denoting an ecological system in which 'every being is no more, or less, than any other. We are all Sisters and Brothers'. In citing this philosophy as a source for deep ecology, Bill Devall and George Sessions emphasize its radical implications: biocentric species equality, feminism, communitarianism and a rejection of the dualistic separation of human beings from nature. All of these principles are antithetical to the conservative values ultimately endorsed by *The Lion King*.[32]

The release of *The Lion King* was accompanied by an unprecedented (at the time) campaign of product franchising, with Disney establishing promotional and licensing tie-ins with Burger King, Mattel, Kodak, Nestlé and Payless Shoesource. The *New York Times* gave an indication of the success of this campaign: 'Sales of Kids Meals by the Burger King chain have tripled since it started giving away the seven plastic figurines with the meals on June 20'.[33] However, although commodity consumption of this kind contradicts the deep ecological rhetoric of the film, it is not necessarily incompatible with the naturalization of consumption that is at the centre of the movie's interest in scientific ecology. Charles Elton's concept of the 'food chain', as Worster points out, conceptualized nature as an economy of producers and consumers, and thereby conformed to the ideological requirements of modern capitalist society. Nature, under the New Ecology, writes Worster, became 'a corporate state, a chain of factories, an assembly line'.[34] Nature in *The Lion King* is similarly an economy in which those at the top of the food chain (lions humanized as middle-class Americans) are justified in their right to consume a nature which is guaranteed to remain endlessly renewing and abundant, as long as their power and authority is not usurped by their undeserving enemies, suitably marked as inferiors in terms of class and ethnicity. The movie thus incorporates and distorts radical deep ecological ideas in order to negotiate a contradiction fundamental to many Hollywood environmentalist movies: namely, an appeal to utopian desires for a re-enchanted nature made by a film which is a luxury commodity produced for global mass consumption.

TWO
The Cinematography of Wilderness Landscapes

'A world with dew still on it': wilderness and the cult of pristine nature

The social construction of 'wilderness' in American culture has mobilized complex assumptions about gender, race, class, and national and regional identity. Andrew Ross alludes to the cultural role played by American cinema in conserving images of wilderness for their symbolic significance. He notes that the idea of transforming the wilderness, once central to Euro-American expansionist ambitions, became in the twentieth century 'the very antithesis of white national identity, now so ideologically dependent upon the conservation of that same wilderness, whether on celluloid, on the Native American reservation, or in the strictly policed territories of the national parks'.[1]

The desire to conserve wilderness is informed by the need of an increasingly urbanized people for landscape that appears pristine and uninhabited by human beings, and therefore becomes a psychological safety valve for anxieties over modern development. The 1964 Wilderness Act, which defines 'wilderness areas' as places where 'the earth and its community of life are untrammeled by man, where man himself is a visitor who does not remain', reflects such desires.[2] For preservationist writers Bill McKibben and Thomas Berry, the American wilderness is an unspoiled Garden of Eden, outside of human history, which should be saved from what they see as the essentially destructive effects of human intervention. Only if wilderness is preserved in its purity can it become a sacred site of spiritual redemption for fallen humanity.[3]

Yet the drawback of this need to conceive of nature as pristine is that it tends to position human beings as fundamentally opposed to, and excluded from, nature. Whereas conservationism evokes this separation to grant permission for human intervention to control and master non-human nature, preservationism uses it to limit or prevent such interventions. William Cronon observes that the cult of wilderness within radical environmentalism, which assumes 'that nature, to be natural, must also be pristine', tends to demonize even agriculture as a violation of nature. Such fantasies, he observes, are an inadequate response to environmental crisis. There is a need instead for an environmental ethic that 'will tell us as much about *using* nature as about *not* using it. The wilderness dualism tends to cast any use as *ab*-use, and thereby denies us a middle ground in which responsible use and non-use might attain some kind of balanced, sustainable relationship.[4]

The cultural construction of wild nature as pristine also serves the ideological purposes of particular groups in American society, who evoke the concept of 'nature', and of what they deem 'natural', to assert that social or political values specific to their own interests are 'natural', and therefore universal and unchangeable. Neil Smith observes that this 'universal conception' of nature is a rhetorical strategy central to the discourses of the ruling groups in American society. 'The overriding function of the universal conception today', he writes, 'is to invest certain social behaviours with the status of natural events by which is meant that these behaviours and characteristics are normal, God-given, unchangeable. Competition, profit, war, private property, sexism, heterosexism, racism, the existence of haves and have nots or of "chiefs and Indians"—the list is endless—all are deemed natural'.[5] As Chapter Three will show, constructions of nature in Hollywood cinema have served as foundational models for social behaviour particularly in the area of gender relations. The rest of this chapter, however, will explore the ways in which cinematographers working in the Hollywood film industry have constructed natural landscapes in keeping with the dominant, popular aesthetic taste for images of pristine nature.

The cinematic construction of natural landscape as pristine is based on an aesthetics of exclusion, that omits from landscape images all signs of human intervention in nature, such as roads, buildings, walls, machinery, telegraph wires and litter. Film aesthetics in this way continue the processes of selection and idealization that shaped earlier forms of landscape imagery, from paintings, dioramas and stereoscopes to popular photography at the start of the twentieth century. Nature tends to be shown only at its pristine best, a tourist gaze from which what is undesirable or ugly is omitted.[6]

Landscape cinematography adds to the list of technical variables associated with still photography (lens size, film stock, natural or artificial lighting, filtration, colour balance and focus) the specifically cinematic variable of camera movement, as well as contextual elements such as editing and soundtrack effects. Certain stylistic choices have become commonplace in the work of landscape cinematographers: aerial tracking shots, widescreen formats, wide angle lenses, sharp focus with a minimum of visible grain, and slow motion.[7] Cinematographers have inherited from still photography the opposition between 'realist' (or 'straight') and 'poetic' (or 'pictorial') styles, and have tended to favour an 'unobtrusive' style which enhances the narrative, contributing elements of tone and mood to the film while subordinating them to the demands of the story.[8]

Phillippe Rousselot's Oscar-winning cinematography for *A River Runs Through It* (1992) constructs the wild river landscapes of Montana in the 1920s in the tradition of what Barbara Novak calls the 'Christianized sublime', a natural landscape iconic of bourgeois democratic values and of regional and national pride.[9] The unspoilt wilderness in the film is a site of moral goodness and spiritual transcendence, in contrast to the corruptions associated with the small town. Significantly, Rousselot composed the natural landscape as pristine by excluding signs of human activity, such as houses, roads, bridges, litter, and river pollution.

In this sense, the movie idealizes and prettifies the natural landscape more than Norman Maclean's novella on which it is based. In this description of the Elkhorn River, for example, Maclean places the Montana landscape within a social and economic context of property ownership which is missing from the film: 'Jim McGregor owned it to its headwaters, and every fence was posted, reading from top to bottom, "No Hunting," "No Fishing", and finally, as an afterthought, "No Trespassing."'[10] The movie, in contrast, removes all signs, fences and other marks of human impacts upon nature, in order to preserve its ideal of nature as pristine and beautiful.

Ironically, in the following passage, Maclean explicitly distances the type of natural landscape valued by the fly-fisherman from that sought by the photographer:

> It was a beautiful stretch of water, either to a fisherman or a photographer, although each would have focused his equipment on a different point. . . . No fish could live out there where the river exploded into the colors and curves that would attract photographers. The fish were in that slow backwash, right in the dirty foam, with the dirt being one of the chief attractions. Part of the speckles would be pollen from pine trees, but most of the dirt was edible insect life that had not survived the waterfall.[11]

Robert Redford's movie omits the dirt and the speckles, emphasized in this passage by Maclean, for a more generalized, easy pastoralism. In the novella, the fish in this stretch of water jump to catch bees attracted by the stench of a dead beaver. But Maclean's evocation of abjection and death in nature is excised from the tourist gaze of the movie.[12] Ecocritic Richard Kerridge observes that litter and pollution tend to be excluded from much nature writing, as unwanted signs of abjection in nature. 'Involved in abjection', he writes, 'is a horror at the materiality of a body engaged, through consumption, expulsion, growth and disintegration, in constant exchange with otherness; abjection is therefore a condition of horror at the self conceived in ecological terms'.[13] According to this analysis, the perpetuation of images of pristine, unspoiled nature works against an understanding of ecological relationships that include human beings as part of natural processes.

The construction of nature as pristine in A *River Runs Through It* also serves an ideological purpose in relation to the film's meditation on the relationship between America's frontier past, childhood and masculine identity. Adulthood is for the narrator Norman (Craig Sheffer) a fall from an original state of innocence associated with the landscapes of childhood, 'a world with dew still on it, more touched by wonder and possibility than any I have since known'. These words are accompanied by a sequence of wide-angle, aerial tracking shots of the rivers and mountains of Montana represented as a beautiful, Christianized sublime. The construction of nature as pristine in this way becomes part of the movie's nostalgia for a lost innocence, both personal and national.

Norman's memories of pristine nature are again evoked when he recalls the time he first asked Jessie (Emily Lloyd) to dance with him, in the speakeasy to which his brother Paul (Brad Pitt) had invited them. The speakeasy is a dark, profane space, associated with both Paul's illegal gambling activities and his erotic dancing with his Native American girlfriend, Mona, whose red dress codes her as fleshly and sensual. In the next scene, Jessie is reading a love letter from Norman, which recalls that night: 'I find I am humming, softly, not to the music, but to something else, some place else: a place remembered, a field of grass where no-one seemed to have been except the deer. And the memory is strengthened by the feeling of you, dancing in my awkward arms'. Norman's words are accompanied by slowly dissolving, lyrical shots of uninhabited landscapes: a forest, then seeds blowing in a field in slow motion, followed by a misty, forested mountainside. These images of wilderness construct nature as a sacred, pure space, analogous to the chaste relationship between Norman and Jessie, and in contrast to the profane space of the speakeasy. Transcendence in nature is also transcendence of the body in this Protestant version of the relationship between white men and nature.

Fly-fishing is represented in the story as an archetypal male sacrament with a feminized, virginal nature, through which a state of Christian grace and transcendence may be achieved. As an act that reaffirms traditional masculine mastery over nature, fly-fishing is also predicated in *A River Runs Through It* on the exclusion of females and 'feminized' men, as potential carriers of corruption. The prostitute Old Rawhide and Norman's brother-in-law Neal, whose effeminacy is doubly guaranteed in the film (he is a tennis-player, and from California), are both violators of the sanctity of pristine nature. Undisciplined, they arrive late for fishing, and get drunk. Old Rawhide, the word 'LOVE' tattooed on her buttocks, is especially signified as the profane, feminine flesh, the antithesis of the sacramental transcendence available to the male heroes.[14]

As in a traditional western, Paul is the individualist outlaw hero, who rebels against the strict code of his father. In one sense, he is punished for his rebellion, when, addicted to gambling, alcohol and brothels, he is eventually killed in a fight over a gambling debt. Yet Paul's rebellious individualism paradoxically gives him superiority in the art of fly-fishing. Taught by his father to cast using a metronome, Paul ultimately rejects such mechanical perfection. He had, as Norman puts it, 'broken free of our father's instruction into a rhythm all his own'. Fly-fishing in the Montana wilderness, Paul is a smiling, blond Adam, a figure of American innocence and exceptionalism. Director Robert Redford's use of slow motion in the fly-fishing scenes signifies an intensification of affect in these moments. When Paul finally lands a large trout after it has dragged him into the river after it, immersion in the sacred waters of the wild river redeems the fallen soul of the imperfect fisherman. Emerging from the water with his trophy, Paul tells his father that in another three years he will be able to 'think like a fish'. For the male, identification with nature gives access to the sacred, and transcends the bodily immanence identified with the town and the female. Accordingly, Norman remembers this moment as one of epiphanic transcendence: 'My brother stood before us, not on a bank of the Big Blackfoot River, but suspended, above the earth, free from all its laws, like a work of art. And I knew just as surely and just as clearly that life is not a work of art, and that the moment could not last'. With this paradox, the movie cuts from the beautiful river at sundown to the police station door, and news of Paul's death in a brawl. The bones in his casting hand, Norman is told, were broken by his assailant. The small town, ominously lit by Satanic smoke and red light, is the profane site of Paul's fall from grace, and the antithesis of the sacred wilderness in which he had been earlier been exalted.

In *A River Runs Through It*, then, the cinematography of natural landscape as pristine serves a wider ideological project concerning the gendered relationship between human beings and nature. In addition, the promotional campaign for the film evoked nostalgia for the cult of pristine nature in order to advocate the ecological restoration of American wild rivers. Robert Redford told *Fly Fishing* magazine: 'This is *our* land, and it is being wiped out along with whole cultures. This movie captures some of what we lost, and it may help to save such rivers as the Blackfoot'.[15] The preview in Bozeman, Montana, was held in aid of the restoration of the Big Blackfoot River, the river on which Maclean's novel was set, but which was so polluted that the movie was actually filmed on the nearby Gallatin, near Livingstone.[16] Further benefit screenings were held in Seattle, in aid of Friends of the Earth and the Washington Wildlife and Recreation Coalition, and in New York, for the Natural Resources Defense Council. Robert F. Kennedy Jr's speech at the latter screening took up the movie's linkage of the American wilderness with Christianity and national identity: 'The movie shows why rivers are important.

. . . they fuel our culture and give us a sense of the divine'.[17] Ironically, however, one immediate effect of the movie's word-of-mouth success was a run on fly-fishing holidays in Montana, putting even more pressure on the ecology of the region.[18] As Gregg Mitman comments in his discussion of Disney's True-Life Adventure films of the 1950s, the 'framing of nature as enter-tainment . . . reinforced a tourist and recreational economy that placed a much greater demand on the very areas that conservationists were trying to protect from the influx of people and the values of consumer society'.[19]

Kinetic landscapes: tracking shots in *On Deadly Ground*

The widespread use of tracking shots in Hollywood landscape cinematography contributes to an aesthetic appreciation of nature as movement. 'The beauty of the American landscape', writes geographer Yi-Fu Tuan, 'seems not designed for the sedentary and the slow-moving—for those who hug the earth'.[20] In this respect, the pleasures of cinematic spectacle parallel other modern technological mediations of nature. David Nye notes that, when the automobile and the aeroplane replaced the railroad as dominant forms of transportation in the twentieth century, so tourists came to demand movement and excitement in their aesthetic experience of nature. 'The railway', he writes, 'promoted static contemplation and the slow accumulation of impressions as one calmly passed a landscape. But the modern tourist wants to penetrate into the site, by car, raft, hang-glider, helicopter, or plane, or if that is impossible, through a film of one or more of these experiences'.[21]

Extended action sequences of this type form some of the central pleasures in movies such as *Deliverance* (1972) and *The River Wild* (1994), where quick cutting and the use of hand-held cameras in close-up simulate the kinetic thrills of a white-water ride. Action cinema of this kind provides consumerist thrills of speed and immediacy, enacting a vicarious sense of mastery over the natural environment. Nature is thereby reproduced for public recreation, rather than for the individualistic, silent contemplation associated with the romantic sublime.

The dual construction of natural landscape as, on the one hand, pristine and uninhabited, and, on the other, an object of kinetic pleasure for the spectator, is a recurrent feature of Hollywood landscape cinematography. Director of photography Ric Waite composed the Alaskan landscape in *On Deadly Ground* (1994) in a way that combines these two tendencies. The opening credit sequence of the film constructs Alaska as a pristine wilderness, whose purity is vulnerable to human intervention. The first shot of the film is a close-up of a bald eagle, looking away from the camera, and then turning towards it. This head movement is synchronized with an ominous drum-beat on the soundtrack: the eagle appears to be have been disturbed by something. After a medium shot of the eagle perched on a rock, there is a cut to a wide-angle, aerial tracking shot that moves over a dark pine forest and a lake towards snow-clad mountains in the distance. Closer shots of the mountain peaks are then intercut with a shot of a polar bear, with the diegetic sound of the animal breathing dubbed over the music. At the end of the credit sequence, the music pauses, and the camera halts on a static shot of what the viewer takes to be either cloud or snow. The silence and stasis are then abruptly broken by the entrance into the frame of a helicopter, noisily penetrating the snowy landscape.

The abruptness of the helicopter's arrival signifies it as a violent intervention into a pristine landscape, and reinforces a sense of the radical separation of human beings from nature. Similarly, the inclusion of the bald eagle and the polar bear at the start of the sequence suggests that the landscape belongs first and foremost to its wildlife, which lives in a harmonious relationship with nature until disturbed by the brutal invasion of human technology. The rest of the movie concerns the efforts of Forrest Taft (Steven Seagal) to reverse the destructive effects of the oil industry in Alaska, whose invasive presence is first announced by the sudden appearance of the helicopter in the opening sequence. At the end of the film, after Taft has dealt with the problem of the evil oil company, the movie returns to the aerial tracking shot over a lake with which it began. A sense of natural equilibrium is thereby re-established, and cutaways to breaching whales, an eagle and a bear suggest that the contaminating presence of modern technology has been expelled and that nature has returned to its pristine state.

Ironically, however, the aerial cinematography used in these sequences also evokes thrills of vicarious mastery over the environment predicated on its technological control. Moreover, the aerial tracking shot is a relatively obtrusive camera movement which draws attention to the processes of cinematic mediation in a way that contradicts the construction of nature in the image itself as pristine and uncontaminated by human presence. In effect, therefore, *On Deadly Ground* has it both ways: constructing nature in traditionally romantic terms as a site of spiritual purity and redemption, and also as a spectacle for thrills of vicarious mastery. This doubleness is typical of the construction of wilderness in Hollywood environmentalist movies.

Theories of landscape and the domination of nature

The implications that this dual representational strategy may have for environmental politics must now be considered. The final section of this chapter will thus consider theories that have speculated on the supposed ideological effects of landscape images.

A common argument in such critical work is that landscape images perpetuate bourgeois relationships to land, endorsing in particular the commodification and domination of nature. As applied to cinematic representation, this claim can be divided into two aspects: firstly, the notion that ideologies of dominance and mastery over nature are inevitably constructed by the camera itself; and secondly, the theory that the perpetuation of those ideologies should be located within specific representational practices, rather than considered an effect of the film medium itself.

The theory of the camera-as-domination has been articulated for still photography in the writings of Susan Sontag. The camera, she argues, by privileging and abstracting the visual sense, creates a sense of detachment from nature, and thereby encourages desires for its mastery and commodification. 'Despite the illusion of giving understanding', she writes, 'what seeing through photographs really invites is an acquisitive relation to the world that nourishes aesthetic awareness and promotes emotional detachment'.[22] A more explicitly environmentalist version of this theory has been developed by philosopher Neil Evernden, who argues that the objectification of nature through photography is a strategy which encourages environmental damage.[23]

In comparison to these theories of still photography, there has been little speculation within theories of the moving image on possible links between cinematic technology and ideologies that lead to the domination or degradation of non-human nature. Although Jean-Louis Comolli and Jean-Louis Baudry asserted in the early 1970s that the cinematic 'apparatus' itself was a means of social 'domination', the concept of domination they explored derived from Louis Althusser's theory of bourgeois class relations, and was characteristically Althusserian in its omission of nature and ecology as relevant issues.[24] Nevertheless, the insights of Sontag and Evernden into still photography have been applied to the moving image by a few critics. Andrew Ross, for example, emphasizes the contradictory role of media representations of non-human nature, describing the camera as 'an embodiment of what ecologists have called the rationalist project of mastering, colonizing, and dominating nature; a project whose historical development now threatens the global ecology with an immediacy that is all the more ironically apparent to us through those very "images of ecology" that have become standard media atrocity fare in recent years'.[25] Jhan Hochman similarly argues that 'film renders viewers separate and superior to film-nature even as it brings them into proximity. Nature becomes, then, prop(erty) and commodity not unconnected to the idea and practice of worldnature as prop for film and property for the larger culture'.[26]

This attribution of an ideology of environmental domination to the very technologies of photography and cinema themselves raises the question as to whether there are potential uses of the camera that avoid such apparently destructive processes. Yet the essentialist and deterministic assumptions of this theoretical approach make positive or oppositional aesthetic strategies difficult to envisage. A second theoretical question therefore arises: whether it is particular formal choices in film production that promote the ideology of domination, rather than the cinematic 'apparatus' itself.[27]

Landscape theorists from painting and photography again provide a critical model which may be extrapolated for film studies. The aerial tracking shot in a film, for example, may be considered the cinematic equivalent of the elevated viewpoint in many nineteenth-century landscape paintings. Albert Boime argues that what he calls the 'magisterial gaze' of these paintings worked in the interests of American Manifest Destiny by reinforcing in the spectator a sense of mastery and possession of the land.[28] The word 'prospect', Denis Cosgrove points out, is linked etymologically to a 'vision of the future', and suggests the land's potential for commodification.[29]

Theorists have argued that landscape cinematography also encourages a possessive and commodifying attitude to the land that serves the interests of property-owning classes. There are two main arguments here. The first one is that films tend to subordinate nature to the centrality of their human dramas, thereby promoting an ideology of anthropocentric mastery and possession of the land. As Aitken and Zonn observe, the 'descriptive and narrative rhythm of film works continually to transform place once more into space as landscapes are decentred to accommodate action and spectacle. . . . Place becomes spectacle, a signifier of the film's subject, a metaphor for the state of mind of the protagonist'.[30] Jhan Hochman brings out the possible implications for environmental politics of this centring of the human:

> Most important for green cultural studies are the capitalist/communist/ technical dominations of worldnature that are informed by a textual nature prone to represent nature unimaginatively and flatly, as a two-dimensional backdrop to the human drama. Material and representational domination is reciprocal and double. Each stands to aggravate or potentiate the other, reifying nature as a realm fit primarily for multiple manipulations and annihilations.[31]

The second argument brings out the environmentalist implications of the tendency of American landscape cinematography to construct nature as pristine and unpeopled, already discussed in this chapter in relation to *A River Runs Through It* and *On Deadly Ground*. In her essay on American television nature documentaries, Karla Armbruster focuses on the ideological effects of images that show animals or plants in shots of pristine nature empty of human beings. These shots, she argues, construct nature as 'a place with no room for human beings, ultimately distancing humans from the non-human nature with which they are biologically and perceptually interconnected, and reinforcing the dominant cultural ideologies responsible for environmental degradation'.[32] Writing on the western movie, Jane Tompkins similarly argues that the cinematography itself reinforces an ideology of domination over nature for the filmic spectator. The 'openness of the space', she writes, 'means that domination can take place virtually through the act of opening one's eyes, through the act, even, of watching a representation on a screen'.[33]

However, the implicit formalism of all of these arguments is problematic. The inclusion in a film or television programme of a particular type of shot does not necessarily guarantee the kind of ideological positioning that these arguments assume, because the meaning of a shot is context-dependent, being produced not only by that shot's relationship with other elements in the film, but also by the filmic spectator's prior knowledge which he or she brings to the viewing process. The notion that meaning is inherent in particular formal practices, such as the individual shots of a film, is derived from the aesthetics of political modernism which assume, as D.N. Rodowick puts it, 'both the essential integrity and self-identity of aesthetic forms, as well as an intrinsic and intractable relation between texts and their spectators, regardless of the historical or social context of that relation'.[34] There is, then, nothing in a shot of an empty landscape that inherently signifies the separation of human beings from nature, or the desire to master and exploit nature. Rather, an interpretation of this kind depends on the spectator's prior knowledge of radical environmentalism, which provides the viewing context within which the image can be interpreted in that way.

Indeed, it is conceivable that an image of a natural landscape as pristine and empty of human presence may encourage in the spectator a desire to keep it that way. Jane Tompkins' own comments later in her chapter on landscape suggest this alternative interpretation, as she notes that the sublimity of the western landscape can evoke feelings of humility in the filmic spectator as much as a desire for mastery:

> The worship of power, the desire for it, and, at the same time, an awe of it bordering on reverence and dread emanate from these panoramic, wide-angle views. . . . The landscape arouses the viewer's desire for, wish to identify with, an object that is over-powering and majestic, an object that draws the viewer ineluctably to itself and crushes him with the thought of its greatness and ineffability.[35]

Although Tompkins' analysis again assumes that meaning resides in the formal aspects of a shot, rather than in its narrative and spectatorial context, the insight that sublime landscape images may not simply perpetuate vicarious feelings of dominance and mastery over nature is a useful one. In the light of such an interpretation, the theoretical linkage of images of pristine nature with spectatorial desires to master and commodify nature appears overly deterministic. As Christine Oravec observes, images of sublime landscapes in American popular culture have served the interests of both preservationism and development, depending on the context of their reception. In nineteenth-century painting, the sublime convention was used both to promote the settlement and development of the land, and, in the case of painters Albert Bierstadt and Thomas Moran, 'to keep a kind of idealized, utopian conception alive that could serve as a model for preserving what little remained'.[36] Rather than simply encouraging a sense of consumerist possession of the land, therefore, the tradition of the sentimental sublime, which has become a commonplace in popular culture from Sierra Club calendars to *On Deadly Ground*, has been useful in preservationist advocacy. An image of a pristine landscape, in which all signs of human intervention have been excluded by the photographer or cinematographer, can suggest the need to preserve that landscape in its undeveloped state.

In any case, the assumption within radical environmentalism that the dualistic separation of human beings from non-human nature (apparently encouraged by landscape art) is a major cause of environmental decline is itself questionable. As Martin W. Lewis argues, the critique of dualism as a cause of environmental damage is founded on a simplistic epistemological determinism. A more direct cause of environmental damage is the application of bad technologies, rather than epistemological attitudes towards nature.[37] Kate Soper similarly argues that 'the human predicament is sufficiently different from that of any other living creature to make it implausible to suppose that metaphysical naturalism is the automatic ally of ecology, dualism (or "humanism") its obvious enemy'.[38]

Both *A River Runs Through It* and *On Deadly Ground* place their images of pristine wilderness in opposition to urban or industrial scenes that connote pollution and the despoliation of nature. In *Deliverance*, the pristine river landscape evoked by Lewis (Burt Reynolds) in the opening sequence is followed by a scene in which the tourists from Atlanta arrive in the countryside and discover rusty cars and other junk outside the mountain men's shack. Kicking away a rusty can, Bobby (Ned Beatty) comments: 'I think this is where everything finishes up. We may just be at the end of the line'. The mood of apocalyptic fatalism in the film arises from a sense that the redemptive purity associated with pristine nature has been destroyed by the forces of modernization. That this emotional and ideological need for pristine nature is central to the construction of masculine identity in American culture has been suggested by the discussion of *A River Runs Through It* earlier in this chapter. The next chapter explores in more detail the complex ways in which encounters with wilderness in Hollywood cinema have provided a context for both the affirmation and the unsettling of gender identities.

THREE

Gender and Encounters with Wilderness

Masculinity: survivalism and sensitivity

Male heroism in environmentalist movies is identified with saving as well as conquering nature. Nevertheless, the wilderness often remains in these fictions a site for male protagonists to recover an essential, authentic masculinity, and thereby to reassert the hegemony of the white male not only over non-human nature, but also over his ethnic, racial and gender subordinates. Women continue to be excluded from, or marginalized within, these narratives of masculine adventure and self-fulfilment. However, as Annette Kolodny has argued, the natural landscape itself is often feminized in American fictions of the wilderness, becoming the object for the projection of Euro-American male fantasies of erotic discovery or rape.[1] In environmentalist movies, nature has also become a space to be protected by the heroic male.

Deliverance (1972) mourns the development of the wilderness for modern industry while putting into question the masculinist paradigm of mastery over a feminized nature. Lewis (Burt Reynolds) identifies the industrialization of the wild river for hydro-electricity as a rape with which the Atlanta men are complicit. 'You push a little more power into Atlanta, a little more air conditioning, for your smug little suburb, and you know what's going to happen: we're gonna rape this whole goddamned landscape. We're gonna rape it'. The Atlanta men's desire to canoe down the river is further sexualized when Lewis promises Bobby: 'You just wait 'til you feel that white water under you'. A feminized nature, Kate Soper comments, is not 'emblematic simply of mastered nature, but also of regrets and guilt over the mastering itself; of nostalgias felt for what is lost or defiled in the very act of possession; and of the emasculating fears inspired by her awesome resistance to seduction'.[2] The guilt, nostalgia and fear involved in male attempts to master nature are precisely what *Deliverance* works through. When the men have successfully negotiated their first set of rapids, Bobby in particular is foolishly complacent about their masculine prowess. 'We beat it, didn't we?', he shouts enthusiastically. 'Didn't we beat that?' Lewis's reply is more cautious: 'You don't beat it. You don't beat this river'. Lewis' display of cautious respect for nature is borne out by the rest of the narrative, as the men's hubristic desire to master nature ends in disastrous failure. Nature does not submit to their attempts at control, but instead exacts retribution, in the guise not only of the wild river which injures Lewis, but also in the sexual violence inflicted on Bobby by the mountain men. The suburban insurance man, who had been openly condescending to the local community, and had bragged insecurely about his past heterosexual conquests, is reduced to a powerless animal as the men sodomize him and force him to squeal like a pig. Linda Ruth Williams observes that 'the initial acts of violence can already be seen as retribution, with the country folk getting revenge for the indignities and injustices wreaked upon them by the townies they now have a chance to victimize'.[3] Moreover, the sexual assault at the centre of the narrative is properly understood, as J.W. Williamson argues, as a 'confrontation between mainstream America and its own hidden potentialities', as the worst impulses of urban America are projected onto the 'hillbillies'.[4]

Despite this punishment for the city men's attempt to master nature, as an environmentalist narrative *Deliverance* is resigned and fatalistic. Instead of protest action, the men are forced to resort to armed survivalism. It is already too late to halt the industrial development of the river valley: the apocalypse has already happened, and mere survivalism is the only option left. However, the movie's attitude towards survivalism is itself ambiguous. When Lewis breaks a leg shooting the rapids, he proves that he is unable to survive without the help of his friends, thereby confirming the judgement made by Drew (Ronny Cox) earlier in the film, that their leader 'can't hack it'. The narrative in this way tests Lewis's masculinist rhetoric of survivalism in conditions of apocalyptic scarcity against the bourgeois, suburban values embodied by Ed (Jon Voigt). The organization man's first encounter with wild nature ended in failure, when he suffered 'buck fever' and was unable to shoot the deer he was tracking. However, with Lewis injured, Ed is forced to assume the role of the group's armed protector, climbing the hillside to kill his pursuer with a bow and arrow. Nature has again become a scene in which traditional ideals of American manhood are tested and reaffirmed, and the natural world brings out in the male a primitive, regenerative violence. Nevertheless, as Linda Ruth Williams comments, the film is nuanced and ambiguous in its construction of masculinity. *Deliverance,* she writes, 'is not so much a film about men looking for an authentic gender identity as a film about masculinity as an agent of change and difference, something which, far from being fixed, immutable, sovereign, can be lost and precariously found, diminished or warped'.[5] Moreover, there is no sense, at the end of the movie, that the men have successfully dominated and mastered nature, nor that such attempts are heroic.

Compared to Lewis and Ed in *Deliverance*, Steven Seagal's character Forrest Taft in *On Deadly Ground* reduces environmentalist awareness to a regressive cult of masculine prowess that lacks the nuances and uncertainties of the earlier movie. However, environmentalist films have shown a trend towards a more sensitive, less brutally macho type, a trend epitomized by Kevin Costner in *Dances With Wolves* (1990), and also evident in the father-figures of recent children-and-animals movies such as *Andre* (1994), *Fly Away Home* (1996) and *Flipper* (1996). Nevertheless, the emergence of what Fred Pfeil calls the 'sensitive guy' in Hollywood movies of the 1980s has often served as an ideological ruse for the preservation of patriarchal power. The point of such narratives for their male protagonists, writes Pfeil, 'is not finally to give up power, but to emerge from a temporary, tonic power shortage as someone more deserving of its possession and more compassionate in its exercise'.[6] Moreover, the redemption of the male in these films is often reinforced by the subordination or demonization of women.

The re-make of *Flipper* (1996) is a case in point. Part of the recent cycle of wild animal movies instigated by the success of *Free Willy* (1993), *Flipper* differs from its 1963 prototype in acknowledging imperfections in the nuclear family. In the earlier movie, Sandy (Luke Halpin) was a twelve-year-old boy whose mother was an idealized housewife, happily cooking and cleaning with no signs of domestic disharmony. In the 1996 version, Sandy (Elijah Wood) is a teenager whose parents have recently divorced. The traumatized adolescent begins to empathize with the dolphin Flipper when it too is separated from its mother, killed by callous fisherman Dirk Moran (Jonathan Banks). Sandy finds in Flipper the unconditional love he lacks in his own family relationships, as contact with benevolent nature provides therapy for the disaffected teenager. Eventually, Sandy gains a social conscience, even giving up the chance to see the Red Hot Chili Peppers in concert to save Flipper's life. Like the orca in the *Free Willy* series, then, the wild dolphin serves the ideological purpose of representing a model of virtuous nature from which conservative moral lessons concerning human gender relations can be learned. For the relationship between human boy and wild animal is represented fundamentally as that of owner and pet, in order to reinforce an Oedipal version of the human family.[7] Moreover, in the subplot that focuses on the relationship between marine biologist Cathy (Chelsea Field) and Sandy's uncle Porter (Paul Hogan), the movie confirms Fred Pfeil's interpretation of the 'sensitive guy' movie as ultimately reinforcing patriarchal values. Having given up her career in marine biology to be near Porter in the Florida Keys,

Cathy's overriding concern throughout the narrative is to get commitment from her man. In the end, Porter buys her the flowers she has always wanted from him, and asks her to marry him. Although Cathy's expertise in marine biology helps to save Flipper's life, the environmentalist movie nevertheless predicates the gaining of sensitivity on the part of the male hero on the positioning of its most important female character in a traditional, supportive role. This hierarchy in the delineation of gender roles raises the broader question of the representation of women in environmentalist movies, a topic which will now be discussed in more detail.

Femininity: action heroines and ecofeminists

Environmentalist movies in which women are the main protagonists tend either to position the female as an action heroine, as in *The River Wild* (1994), or to evoke the spiritual aspects of ecofeminism, as in *FernGully: The Last Rainforest* (1992) and *Pocahontas* (1995), the latter of which will be discussed in the next chapter with regard to the figure of the ecological Indian.

The River Wild is a partial revision of the masculinist wilderness survival narrative, and draws on the figure of the action heroine that emerged in Hollywood cinema in the 1980s in movies such as *Aliens* (1986) and *Terminator 2: Judgement Day* (1991).[8] Gail (Meryl Streep) is both action heroine and super-mother; the movie thereby appropriating the new figure of feminist agency to reinforce the more traditional female role.

When Gail and her family are kidnapped by a gang of armed robbers, the wilderness becomes a proving ground for female heroism, as her skill in white-water rafting saves her middle-class family from the threat embodied by the criminal 'underclass'. To an extent, *The River Wild* thus positions its female protagonist in a role traditionally assumed by male characters. Gail is a woman who knows nature, knowing where to fish and how to negotiate rapids. A female version of the American Adam, Gail's respect for nature is guaranteed by her knowledge of, and proximity to, Native American culture: she tells her son Roarke (Joseph Mazzello) about the vision-quest, and is friends with the Native American ranger Johnny (Benjamin Bratt). As an action heroine, Gail takes charge of the raft and successfully leads it through a set of dangerous rapids, thereby fulfilling her own self-actualization philosophy ('be the best you can be') and demonstrating to the leader of the criminal gang, Wade (Kevin Bacon), that she does have the 'guts' after all. In this way, the movie partly questions the excessive machismo associated with the male action-adventure movie. Gail and her husband defeat their enemies through superior intelligence rather than brute force, outwitting rather than out-fighting them. Nevertheless, when Gail is forced to kill Wade with his own gun, she finally assumes the traditionally masculine role of reluctant gunfighter, as the

American wilderness imparts its familiar lesson, now to women as well as to men, that violence is justified in self-defence and may be necessary for survival.

However ambiguous, the movie's construction of Gail as a figure of liberal feminist agency is ultimately compromised not only by its emphasis on her maternal role, but also by its overriding concern to rehabilitate the authority of the father, to the extent that the heroism of the male eventually displaces that of the female. At the start of the narrative, Tom (David Strathairn), the architect father, is working too hard and neglecting his wife and children. Indeed, Gail accuses her husband of working long hours to avoid contact with his family. Moreover, the failing father is initially the object of comedy in the film, incompetently falling into the river, and even being disobeyed by the family dog. By the end of the movie, however, Tom successfully learns to use Native American pictographs and smoke signals to save his family from the criminal gang. He has mastered the wilderness, and thereby regains the respect of his wife and son.

Significantly, despite Gail's contribution to saving her family, it is Tom who is celebrated as the movie's ultimate hero. As in the 1996 version of *Flipper*, then, the rehabilitation of male authority is reinforced by the subordination of the female. At the crisis point in their relationship, Tom shifts the blame for their failing marriage from himself to Gail, accusing her of setting him 'high expectations to live up to'. 'For once in your life', he tells her, 'don't be first'. The end of the film validates the hierarchy implied by these words when Roarke gives more credit to his father for saving their lives than he gives to his mother. He was not scared of the river, he says, because he 'knew mom could handle that'. However, when asked, 'What did your dad do?', he replies proudly: 'He saved our lives'. The father, not the mother, is now first. Gail's competence has simply been taken for granted: it is the father who has proved himself, and therefore deserves the most credit. So the movie ends by detracting from the heroism of its female protagonist, celebrating instead her reconciliation with her husband, and the discovery of his heroic masculinity in the wilderness.

Although she wears a 'Save the Earth' T-shirt at the start of her holiday, and is conversant with the Native American vision-quest, Gail in *The River Wild* is not developed as an ecofeminist heroine. The children's animation movies *FernGully: The Last Rainforest* and *Pocahontas*, on the other hand, have appropriated some of the more spiritual forms of ecofeminism, in order to associate their female protagonists with a nurturing attitude to non-human nature, within narratives that centre on female self-discovery and personal choice.

The term 'ecofeminism' denotes a wide range of different ecological feminist philosophies, including liberal, Marxist, cultural and socialist. Significantly, it is 'cultural ecofeminism' on which *FernGully: The Last Rainforest* and

Pocahontas draw. Cultural ecofeminism views the modern Enlightenment project of scientific and technological rationality as fundamentally a masculine will to power over both women and non-human nature. Cultural ecofeminists thus oppose what they see as 'masculine' concepts, such as mastery, domination and hierarchy, while celebrating what they identify as 'feminine' values, such as nurturance, empathy and pacifism. Moreover, they posit holistic philosophy as the healing alternative to what they understand as the 'reductionism' of scientific and technological rationality.

Compared to socialist ecofeminism, in which the identification of women with nature, as dual victims of patriarchal oppression, has led to the development of a politics of social justice, cultural feminism has developed a less overtly political form that focuses on New Age spiritualities and pre-Christian religions that personify and worship a feminized nature. Starhawk, for example, asserts that ancient goddess cults allowed human beings to live in harmony with nature, and calls for the re-enchantment of nature.[9]

Unsurprisingly, it is this spiritual, less political form of ecofeminism that has been drawn upon by popular environmentalist movies. *FernGully: The Last Rainforest*, an Australian co-production with Twentieth Century Fox, celebrates a feminized, pagan nature, ruled by Magi Lune, a Great Goddess-figure with long grey hair and flowing skirts, as a harmonious alternative to the hard-edged, masculine technologies of industrial development. For critic Erik Davis, the wise old matriarch, 'with one wave of her hand almost vaporizes the evil hag archetype that Disney and the Brothers Grimm have been ramming into the cultural subconscious for centuries'. [10] Magi Lune embodies a vision of nature that is holistic and animistic. 'Remember', she tells the fairy Crysta, 'all the magic of creation exists within a single tiny seed'.

The narrative of *FernGully* is centred on the self-development of Crysta, who is exhorted by Magi Lune to find 'inside herself' the solution to the destruction of the rain forest. Crysta also plays the role of redemptive female who saves the blond, handsome logger Zak from destroying the rain forest. In the opening sequence of the film, Magi Lune recounts to Crysta the story of the fall of the rain forest from its original state of perfection:

> Our world was much larger then. The forest went on forever. We tree spirits nurtured the harmony of all living things. But our closest friends were humans. Then, as sometimes happens, the balance of nature shifted, and Hexxus, the very spirit of destruction, rose up from the bowels of the earth and rained down his poison. The forest was nearly destroyed. Many lives were lost, and the humans fled in fear, never to return. Most think they didn't survive.

Magi Lune imprisoned Hexxus inside an enchanted tree, thereby allowing the forest to return to its former state of purity.

The rain forest is represented in this speech as pristine, balanced, harmoniously interconnected and benign. As Magi Lune explains to Crysta: 'Everything in our world is connected by the delicate strands of the web of life, which is balanced between forces of destruction and the magic forces of creation'. To denote the 'forces of destruction', the opening sequence visually identifies Hexxus with a volcano erupting into the rain forest. Nature, apparently, has no place for such forces of imbalance, chaos and catastrophic destruction. It is, rather, a place of pristine health. Hexxus cannot survive in the newly restored FernGully, Magi Lune tells Crysta, because 'There are no poisons here on which he can feed'. The movie's rhetoric of interconnectedness thus ignores the possibility that destruction and creation may be mutually interdependent. In a similar way to the fire that destroys the forest in *Bambi*, the forces of destruction are seen as unnatural, rather than as parts of nature necessary for ecological renewal. There can be no role for instability in this preservationist vision of nature as timeless, static balance.

The plot of the movie involves the threat to the rain forest when Hexxus is freed from his tree prison by the Leveller, a logging machine. Hexxus uses the Leveller to begin to destroy the rain forest. As in *Bambi*, then, *FernGully*'s preservationist attitude to nature views human intervention in natural processes as essentially destructive and unnatural. When Crysta discovers the trees cut down by the Leveller, Magi Lune informs her that 'a force outside of nature did this'. In particular, the movie demonizes industrial technology as the cause of environmental degradation. Hexxus, the spirit of destruction, resembles a brown oil slick, while the saw mill in the forest is depicted as a red infernal machine, making a cacophonous noise. This is part of a universal technophobia expressed in the movie: when Crysta asks Zak what a 'machine' is, he replies, a 'thing for cutting down trees'. The movie thus posits all machines as bad, and logging as an essentially destructive act.

FernGully represents instead the fantasy of a re-enchanted, animistic nature familiar to cultural ecofeminism. 'Don't you miss talking to the forest?', Crysta asks the human interloper Zak. 'Can't you feel its pain?', she remonstrates with him, after he carves her name on a tree. This paganized view of nature is part of the movie's nostalgia for the 'countercultural' values of the late 1960s, an allegiance reinforced by the psychedelic visual style of the animation itself. 'What's a job?', Crysta asks Zak at one point. Batty (Robin Williams), a manic bat victimized by vivisection experiments, makes the authoritative connection between urbanization, consumerism and the destruction of nature. 'There goes the neighbourhood', he jokes, before tracing a history of technological development as a decline into consumerist banality. 'First thing, all these trees go. Then come your highways, then come your shopping malls, and your parking lots and your convenience stores, and then come . . . "price check on prune juice, Bob, price check on prune juice" . . . '

Yet this satirical denunciation of consumerism is contradicted by the movie's celebration of commodified youth cultural style. Zak teaches Crysta hip teenage language ('cool', 'bodacious babe', 'bad') reminiscent of *Bill and Ted's Excellent Adventure* (1989) or *Wayne's World* (1992). 'We're communicating now', he tells her. Later, he teaches the fairies to dance to 'Land of a Thousand Dances', and gives his personal stereo to Crysta's friend Pip, which the fairy ultimately prefers to his own Pan-pipes.

In the utopian space of *FernGully*, then, technologies of production are demonized, while technologies associated with leisure consumption are naturalized as benign and desirable. At the centre of this ambivalence lie class and racial anxieties. Like *The Lion King*, *FernGully* evokes an ecological rhetoric of nature to put in place a conservative social and political ideology. Zak is the blonde, square-jawed male hero, in a movie in which the Australian rain forest, like the Serengeti in *The Lion King*, is represented as a National Park, unaccountably empty of its darker-skinned, aboriginal inhabitants. The fairies, 'the guardians and the healers of the forest', are instead white, middle-class Americans. Moreover, threats to the natural order of the forest are coded as Other. The goanna who tries to eat Zak is voiced by African-American rapper Tone-Loc. If racial anxieties are evoked in this way, class anxieties are also connoted by the proletarian operators of the Leveller, who are stereotyped as greedy and lazy.

At the end of the movie, the pristine rain forest is saved from the threat of logging in a two-fold resolution, which reconciles the technological and the mystical: Zak disarms the Leveller by turning off its engine, while Crysta turns Hexxus into a tree by throwing a magic seed in his mouth. Zak then plants the seed from which the forest is regenerated, not through violence, but through a New Age combination of magic and populist American psycho-therapeutics. Zak is redeemed through contact with feminized nature, and returns to human society with an urgent message: 'Guys, things have gotta change'.

Ultimately, the pious ecological sentiments articulated by *FernGully* can be seen as a greenwashing strategy for the Australian conglomerate FAI Insurances Ltd which produced it, a company which *Fortune* magazine pointed out is involved in the coal business.[11] Crucially, the mystical, depoliticized stance towards environmentalism taken by the movie conveniently evades the contradictions and problems involved in the company's position as sponsor of a self-consciously 'green' movie. Moreover, as social ecologist Janet Biehl argues, the more spiritual versions of ecofeminism have themselves been commodified in a way that serves the demands of a globalized consumer culture. The breast of the goddess, as she puts it, 'can be a symbol of sustenance—or it can be a symbol of dependency on a consumer culture'. [12]

Moreover, the recourse to feminized magic, endorsed by *FernGully*, is also limited as a strategy for environmental restoration. Mythopoeic approaches to the world, as Biehl points out, cannot be relied on for explanations of how non-human nature works. Magic, she writes, 'by its very definition in terms of what we know today ignores any real relationships between causes and effects. It seeks effects essentially without any causes at all'. [13] The explanatory powers of science and reason may be partial and limited, but are nevertheless real, and form a much more effective basis for environmental politics than mythopoeia and spiritualism. Kate Soper similarly finds the concept of the 're-enchantment of nature' of doubtful value for an effective environmental politics. Ruling elites, she observes, have used the divine Goddess figure as a theological caution against Promethean desires not only to master nature but also to throw off human oppression. She concludes that concern for nature, rather than mystical awe and reverence for it, is more likely to lead to effective environmentalist strategies. 'The sense of rupture and distance that has been encouraged by secular rationality may be better overcome', she argues, 'not by worshipping this "other" to humanity, but through a process of re-sensitization to our combined separation from it and dependence on it'. [14]

Such defences of rationality and science clearly buck the trend not only of many forms of radical environmentalism, such as deep ecology and ecofeminism, but also of many Hollywood environmentalist movies. As the next chapter will show, the figure of the ecological Indian is another commonplace in such movies that evokes romantic desires for the re-enchantment of nature.

FOUR

Ecological Indians and the Myth of Primal Purity

The myth of pristine wilderness is complemented by that of the deep ecological American Indian living a life of primal purity. This myth constructs pre-industrial peoples as a basis in nature for the claims of radical environmentalism, by assuming that they embody alternative lifestyles which hold the key to a more harmonious relationship with nature. Environmental philosopher Neil Evernden explains the connection clearly: 'Anyone seeking the truth, the eternal standards by which humans ought to live, would have to inquire which standards are given by nature. Hence the widespread interest in "primitives", who are often presumed to be living by those primitive standards'.[1]

The romanticization of the American Indian depends on a conceptualization of nature as benevolent, because the ideal of humanity living close to nature is unappealing if nature is considered a place of savagery and violence. Accordingly, as Martin W. Lewis observes, radical environmentalists have embraced a model of nature that 'glorifies the harmonious functioning of undisturbed ecosystems and that considers co-operation among individuals and among species far more common than competition'.[2] However, many anthropologists now contest the notion that small-scale hunter-gatherer societies lived in harmony with nature, and used natural resources in sustainable ways with no significant impact on their environment. Such societies, for example, built earthworks and set fires which radically changed the landscape, and exhausted soils through poor irrigation methods. Even the supposedly pristine Eden of the rain forests are in part the anthropogenic product of Paleo-Indians burning trees and scattering seeds. Moreover, some societies killed large, unsustainable numbers of game, and not only for subsistence purposes. Nor did all tribal groups possess a conservationist ethic.[3]

That a society based on non-industrial technologies may damage the environment less than one based on industrialism is a plausible hypothesis. But the idealization of Native cultures as living in a pristine, unfallen Eden is more a product of white liberal guilt and wishful thinking than of such consid-erations. Nevertheless, the notion that hunter-gatherer societies had little significant effect on ecological conditions has been a commonplace in some environmentalist and academic-left writings on human ancestry since the 1960s.[4] The prelapsarian Noble Savage serves as a figure of moral absolution for the bad conscience of the industrial world, attesting to the continuing need on the part of some white liberal Americans to construct a pure, original Other that embodies absent values of authenticity and community, and thereby transcends the alienations of modernity.

Unsurprisingly, the ecological Native American has also become a commonplace figure in Hollywood movies, not only those set in the historical past, such as *The White Dawn* (1974), *Dances With Wolves* (1990) (which will be discussed in Chapter Six), *The Last of the Mohicans* (1992), *White Fang II* (1994) and *Pocahontas* (1995), but also in contemporary stories such as *The Savage Innocents* (1960), *The Bears and I* (1974), *Never Cry Wolf* (1983), *Thunderheart* (1992), *On Deadly Ground* (1994), the *Free Willy* series (1993-7) and *Alaska* (1996). These films perpetuate several commonplace notions about the ecological Indian, as the following sections will demonstrate.

Indian land and National Parks: *The Bears and I*

Disney's *The Bears and I* (1974) adds to Robert Frank Leslie's account of his life with three bear cubs in the Canadian north woods a subplot involving a conflict over land use between a Native American tribe and the federal government. In Leslie's book, a coalition of loggers, miners and trappers successfully lobby the federal government to prevent the designation of the Babine Lake area as a nature reserve or park, with the local Indian population divided on the issue.[5] In the absence of park status, sports hunters eventually wound and kill two of the orphaned bears which Leslie had been looking after with a view to returning to the wild. (The relationship between human being and wild animal in the film will be discussed in Chapter Eight).

In contrast to Leslie's book, the movie version of *The Bears and I* ultimately validates, rather than criticizes, federal policies towards the Indians, by contriving a reconciliation between Indian and white interests through the mediation of outsider Bob Leslie (Patrick Wayne), who is identified as a man of goodwill friendly towards the Indians. When a government official announces to the Indian tribe that it faces eviction onto a reserve so that the government can turn the land into a National Park, he presents an argument that the subsequent course of the narrative will endorse. 'We know you're angry and we understand why', he tells the tribal gathering. 'But the point is, that this lake, and all the area around here, is government property. Everybody in this country, including you, have a right to use it and enjoy it. That's why the decision was made to develop this into a National Park'.

When asked by the Commissioner to discuss the matter with the tribal council, Bob Leslie goes to the heart of the unequal power relationships implied by the federal relocation policy. 'But if the Indians have no choice', he asks the Commissioner, 'what is there to talk about?' Nevertheless, Bob stands for moderation and reconciliation between the parties, and he eventually manages to convince the tribal chief Peter A-Tas-Ka-Nay (Chief Dan George) of the advantages of co-operating with the federal government. Hence the resolution of the conflict offered by the movie finesses the issue of coercion of the Indians by the federal government: the tribal chief agrees to Bob's compromise proposal, whereby the tribespeople are all appointed park rangers, and thereby allowed to continue living within the National Park boundaries.

The Bears and I in this way endorses both a non-violent solution to the conflict over land use, and the benevolent paternalism of the federal government. Indeed, Sam Eagle Speaker (Valentin De Vargas), an advocate of violent resistance to white interests, is depicted negatively as a drunken rabble-rouser, mercenary and extreme. Chief A-Tas-Ka-Nay, in contrast, is a man of peace. When Bob tells the Chief that the Indians 'can no longer live by the old way', the latter at first maintains his willingness to die rather than move from his home. In the end, however, the Chief accepts the Deputy Ranger's badge as a token that 'from this day forward, we would live in peace together'.

The Indians in the movie are given responsibility by the federal government for animal control in the National Park. 'We control the animals by fishing and hunting,' the Chief tells Bob Leslie, 'but only for our food'. The resolution of the film is thus an ideological endorsement of contemporary federal policies that have since the 1970s encouraged traditional usage rights for Native peoples in some National Parks. Significantly, however, the movie does not extend this endorsement of Indian autonomy from the cultural to the political sphere. The political sovereignty argued for by Sam Eagle Speaker is no longer an issue at the end of the film, as the Chief blithely accepts the paternalism of the federal government. When Bob comments that nothing has changed, the Chief corrects him: 'Oh yes, something has changed. They pay us now, they build us houses. Sometimes the ways of the white man's difficult to understand. But it's all the will of the Great Spirit'. The movie ends, then, with the Indians assimilated to modern notions of progress and a wage economy, and ignoring the question of political sovereignty over the land. The ecological conscience of the Native American is thus shown to be fully reconcilable with official federal policies towards the land.[6]

Anachronism and the Inupiat: *On Deadly Ground*

The production notes to *On Deadly Ground* make self-satisfied statements vaunting the movie's authenticity, citing the band of technical advisers brought in to provide 'extensive research into Native Alaskan heritage, architecture and design'. They continue: 'The attention to detail was exacting, from implements of hunting, fishing and tools, to shelter and clothing. The filmmakers also paid strict attention to religious, natural and spiritual elements of the Native Alaskans. Even the language spoken by the Inuit in the picture is absolutely authentic . . . '.[7]

In fact, the movie's claim to authenticity is exaggerated. One of the 'cultural advisers' brought in by the producers, actor Apanguluk Charlie Kairaiuak, later criticized the film for being 'loaded with inaccuracies and misrepresentations of Alaskan native culture'. He pointed out, for example, that the native village in the film is set in Inupiat territory along the Arctic coast, but the actors speak the Yup'ik language of tribes living near the Bering Sea. As in *Dances With Wolves*, then, subtitles are used to signify a linguistic authenticity that is in fact spurious.[8] However, some of Kairaiuak's advice was taken: he was able to ensure that the film-makers built sod houses in their reconstruction of a 'primitive village', abandoning their original intention of building igloos out of snow and ice. Yet the fact that most Native Alaskans now live in modern houses and cabins was completely ignored. The movie also shows them wearing furs and skins, rather than clothes made of modern materials, and carrying spears, rather than the high-powered rifles used by contemporary hunters.

Authenticity for the film-makers, then, was tied to a mythic conception of traditional Native Alaskan life, and a refusal to represent Native peoples as full members of modern American society. Indeed, the production notes for the movie admit that the intention of the film-makers was to recreate 'a very traditional culture which rarely exists anymore, but reflects the old way of Native Alaskan life'. [9] The film thus refuses to explore the Inupiat as a contemporary people, but instead fixes them in an idealized past. The movie thereby largely ignores the complex contradictions in Native Alaskan society as it attempts to maintain or recreate 'traditional' cultural elements in the contemporary United States. The notion of primal purity, as Martin W. Lewis argues, obstructs the possibility of Native peoples being accepted as full members of the contemporary world, by relegating them to a 'cultural zoo'.[10] *On Deadly Ground*, by idealizing Native Alaskans, effectively robs them of their own identity, turning them into objects onto which white anxieties and desires are projected. The movie thus confirms the comments of environmental historian O. Douglas Schwarz on what he calls the 'myopia' prevalent amongst some American environmentalists: 'We admire Indians so long as they appear to remain what we imagine and desire them to be: ecologically noble savages symbolizing a better way of life than we ourselves find it practical to live. We respect their traditions so long as they fit into our preconceived notions of what those traditions should be'.[11] Like Kevin Costner in *Dances With Wolves*, then, Steven Seagal, as director and star, cast himself within his own movie as a Messianic saviour and leader of an oppressed non-white people.

On Deadly Ground also repeats the old Hollywood tendency to homogenize Native American cultures. Despite its claims for authenticity, the movie creates a fictive culture that is a combination of features taken from Native peoples from all over the United States, both past and present, including a generic Mother Earth spirituality, jewellery and horseriding from the Great Plains, and raven mythology from the Tlingits of Southeast Alaska. When Forrest Taft asks his female companion Masu (Joan Chen) if she knows how to ride a horse, she replies: 'Of course, I'm a Native American'. That the Inupiat are mainly played by Japanese and Korean American actors merely perpetuates the casting policies familiar throughout the history of Hollywood portrayals of Native American peoples.

Ultimately, the positioning by *On Deadly Ground* of the Native Alaskans as primitives justifies the paternalistic leadership of the white American male in fighting for their interests against the machinations of the corrupt oil company. (The movie's representation of the Alaskan oil industry will be discussed at greater length in Chapter Eleven.) The movie is thus contradictory in its attitude towards Native American spiritual beliefs. Although Forrest Taft undergoes a form of spiritual rebirth in a shamanic initiation ceremony, acquiring the power of his bear spirit guide, at the crucial moment in his battle against the oil company he renounces the pacifist spirituality that the movie attributes to the Inupiat, and decides to blow up the oil rig in order (he argues, somewhat implausibly) to prevent a greater environmental catastrophe. In doing so, he overrides, in the typical manner of the western hero, the desire of his female companion to seek a resolution of the conflict through peaceful means. 'Do you really think that this hocus-pocus spirit stuff is going to help us now?', he asks, before proceeding to open up his secret weapons store. At this crucial moment, the violent, technological imperative appropriate to the action-adventure genre takes precedence over Native Alaskan spiritual beliefs.

However, an ironic interpretation of this scene is possible, in which the movie verges on self-parody, self-reflectively aware of the melodramatic implausibility of the resolution it proffers. 'I didn't want to resort to violence', Forrest tells Masu, pausing archly in what could be taken as a parody of the reluctant gunfighter of the classic western. As he opens the door to his weapons store, a framed picture of a bald eagle is promptly replaced, on the other side of the door, by a poster of a man firing a gun. Having made his decision to fight, Forrest proceeds to chose his weapons, again in a manner that verges on self-parody: 'Gimme this FA shotgun with all the magazines and ammo you got, gimme one M-14, a couple of 25s and the SSG, and . . . I think that'll do it'.

That this crucial scene is open to ironic interpretation ultimately suggests that the movie is trying to have it both ways, adding a note of uncertainty to the white male's assumption of power over his Native Alaskan allies. This ambivalence also extends to Forrest's dismissal of Native Alaskan spirituality. For at the end of the movie, after he has successfully saved the Native people from the corrupt oil company, the raven that embodies the spirit of Masu's shaman father, murdered by the oil company, is seen flying off into the mountains. Native spirituality, earlier dismissed by Forrest as an ineffectual response to the 'cold, hard reality of this world', thus returns in the narrative's closing moments, as the movie ends by reasserting its ideological need for the figure of the ecological Native American.

The demand of the action-adventure movie for the fetishization of high-speed, state-of-the-art technologies under the control of a white male leader produces a moment of comic incongruity in the movie, when the Inupiat's traditional wooden sledge is shown to reveal a powerful Yamaha snowmobile underneath, 'for emergencies', as Masu puts it. In order to conform to the expectations of the action genre, the movie thus belatedly recognizes a technology that the Inupiat have been using since the 1960s, not 'for emergencies' only, but routinely. At this point in the film, the existence of the Inupiat as a contemporary people breaks into the narrative, only to be repressed again in the return of the Native Americans to their more usual role in Hollywood cinema, as a natural ground of primitive innocence on which can be enacted the hegemony of the white American male hero.

Disney's sacred hoop: *Pocahontas*

Disney's *Pocahontas* (1995) combines references to deep ecology, ecofeminism and the ecological Native American. The movie blames the destruction of the American environment on the value system of the English colonialists, who 'prowl the earth like ravenous wolves, consuming everything in their path', as the Indian shaman puts it (even though the notion of the 'ravenous' wolf is derived, not from Native American traditions, but from European attitudes to the animal).[12] In the name of freedom, prosperity and adventure, the Virginia Company cuts down trees, digs up the earth, and starts wars, enacting thereby an imperialist project to dominate the earth that the movie associates with a paranoiac, masculine will to power. As the rapacious English Governor puts it: 'A man's not a man, unless he knows how to shoot'.

In contrast to the English invaders, the native Powhatans are represented in the film as an innocent and nature-venerating people. At one point,

Pocahontas and her friend are shown gathering corn in wooden bowls. Apart from this, the tribe appear to exist without technology, living a joyful, pastoral existence in harmony with a natural world that is safely benign and domesticated. Historian Simon Schama points out that, despite the credits mentioning a Native American consultant from the Mattaponi tribe, the movie makes no real attempt at authenticity: the Algonquin Powhatan nature religion is represented inaccurately, their music is Western in structure, while the nature myths in the movie (talking trees and the river of life) are Western in origin. The Indians are thus represented as a condescending 'positive' image that does not allow for the flaws or complexities attendant on being treated as fully human.[13]

Pocahontas views the European value system as wholly inferior to that of the Native Americans. When John Smith tells Pocahontas about gold ('it's yellow, comes out of the ground, it's really valuable'), she thinks he is talking about sweetcorn. Smith's patronizing attitude ('There's so much we can teach you', he boasts, 'we've improved the lives of savages all over the world') elicits an angry rejoinder from Pocahontas, who rejects the term 'savage'. 'Still I cannot see', she replies, 'if the savage one is me'. In response to his arrogant, colonialist attitude, she sings 'Colors of the Wind', a song which calls for harmony between 'white' and 'copper-skinned' humans beings, and between human beings and the rest of non-human nature.

The song accuses the English of dis-enchanting nature. They think of land only in terms of ownership, assuming that the earth is a dead thing they can claim, instead of an animistic world where every rock and tree is alive with spirit. Pocahontas goes on to accuse Smith of only accepting people who look and think like he does. During this verse, she stops him from shooting at a bear, before the couple discover that the bear has a mate and cubs. The typical Disney image of a human nuclear family projected onto non-human nature here becomes a deep ecological lesson in the kinship of all living creatures. Pocahontas evokes her interdependence with her 'friends' the other animals (the wolf, the bobcat, the heron and the otter) as nature is constructed anthropomorphically as arcadian, abundant and benevolent. However, this pious invocation of a biocentric notion of species-equality is contradicted by the treatment of animals in the rest of the movie, which continues the Disney tradition of using animals, in this case Meeko the raccoon and Flit the hummingbird, solely as comic relief, peripheral to the human interest drama.

The reference in 'Colors of the Wind' to the 'hoop' in which the universe is contained invokes the Plains Indian notion of the 'sacred hoop', a holistic conception of life as an endless dynamic unity within which all things are connected. This ecological notion is startlingly visualized as a close-up of Pocahontas and John Smith zooms out to reveal them as a reflection in the eye of a bald eagle. The imagery of the hoop is reminiscent of a key work of ecofeminism, Paula Gunn Allen's *The Sacred Hoop* (1986).[14] In transposing the image of the sacred hoop from the Plains Indians onto the Powhatans, however, the Disney movie engages again in the homogenization of Native American cultures typical of Hollywood cinema.

Pocahontas ultimately displaces its concerns with ecology, stated most explicitly in 'Colors of the Wind', into a narrative of female self-actualization, which essentializes femininity in traditional terms as a life-affirming flux. Pocahontas, she says, goes wherever the wind takes her. Moreover, when her father advises her to be steady like the river, she knows that the river is not steady, but always changing and flowing. When she paddles her canoe past an otter building a dam, it reminds her of her 'sturdy' husband-to-be 'building sturdy walls', but incapable of dreaming of the unexpected. In contrast, Pocahontas is an adventurous, risk-taking heroine, taking on the rapids, and intent on discovering her true path in life. The path that she finally discovers is that of her mother's role as peace-maker and source of wisdom for the community. At the end of the film, then, Pocahontas must relinquish her desire for John Smith to do her duty for her people.

The racial implications of this version of the Pocahontas narrative thus tend to dilute the call for unity between 'white' and 'copper-skinned' peoples articulated by 'Colors of the Wind'. For when Pocahontas rejects her dull but worthy Powhatan suitor, the warrior-hero Kocoum, and falls in love at first sight with the blond, square-jawed white hero John Smith, her choice perpetuates the old colonialist stereotype of the woman of colour recognizing the supposed superiority of the white European male. Moreover, the ending of the movie is cautious and evasive in its racial implications. In his study of Pocahontas narratives, Robert Tilton shows how fear of miscegenation in nineteenth-century America led to a preponderance of narratives focusing on Pocahontas's relationship with John Smith, as opposed to the man whom the historical woman actually married, John Rolfe. Choosing the John Smith narrative was a way of avoiding the dangerous issues of inter-racial sexuality and marriage. Instead, the story of John Smith and Pocahontas is a narrative of what Tilton calls 'conspicuous non-consummation', in which a sexual relationship between the races could be anticipated, but finally prevented.[15]

The narrative resolution of Disney's *Pocahontas* perpetuates this disavowal of racial mixing. After Kocoum has been killed by the English, Pocahontas saves John Smith's life. He then reciprocates by saving her father from death at the hands of the English Governor. Yet Smith's actions ultimately prevent the possibility of the two lovers being together in the future: injured as he protects Pocahontas's father by throwing himself in front of the English Governor's bullet, Smith must return to England for medical treatment. As Pocahontas's new-found sense of duty to her community prevents her from going with him and following her desire, inter-racial love remains unconsummated, and the two races return to their separate spheres. The English colonialists return to England, leaving the Indians to their Eden. So, despite the movie's message of racial tolerance and mutual understanding overcoming distrust, the ending plays safe. The promise of the inter-racial kiss earlier in the movie remains unfulfilled, and the lovers' union is sublimated into the spiritual realm in the final image of the windblown leaves.

As a discourse on ecology, the ending of *Pocahontas* reinforces the preservationist ideology typical of Disney movies: nature, having been violated by the white invaders, is best left alone. Humans beings, as so often in Disney, are ultimately excluded from a nature that properly belongs either to animals like Bambi or Simba, or here, in the submerged racialist discourse of *Pocahontas*, to a primal Native American people still romantically associated with nature rather than with culture and history. In a final irony, the balance of nature is only restored in *Pocahontas* by the omission of the story's historical referent: the subsequent history of conquest in the founding of Virginia by the English. Moreover, if the pious deep ecological sentiments of *Pocahontas* are placed within the context of the poor wages paid to workers in Haiti employed to sew Pocahontas pyjamas, then the movie's deep ecological rhetoric, like that of *The Lion King* discussed in Chapter One, becomes another example of a greenwashing alibi, this time for the very forces of global capitalism the movie overtly criticizes.[16]

Resisting colonialism: *Thunderheart*

Thunderheart (1992) is a thriller based on events in 1975 at Pine Ridge Reservation, South Dakota. Director Michael Apted also made the documentary *Incident at Oglala* (1991), about imprisoned activist Leonard Peltier, allegedly framed for the murder of two FBI agents.[17]

Compared to *On Deadly Ground*, the movie makes a good attempt at representing Native Americans as a contemporary people engaged in a political struggle against internal colonialist oppression. The story revolves around an issue of environmental justice: a conspiracy involving FBI agent Frank Coutelle (Sam Shepard) and Jack Milton, the pro-government tribal president, to conceal, under the guise of National Security, the secret strip mining for uranium on Sioux reservation land, a project undertaken despite the people voting against it. The mining is contaminating the river, and, if continued, will kill off the Sioux living in the reservation. The denouement reveals that Coutelle and Milton were behind both the murder of Leo Fast Elk, who had discovered the truth about the uranium mining, and the framing of Jimmy Looks Twice, a militant activist with the Aboriginal Rights Movement, to discredit his organisation.

The movie's thriller structure delays revelation of the environmental conspiracy from both the protagonist, the mixed race FBI agent Ray (Val Kilmer), and the viewer until near the end. This structure thereby displaces attention away from the issue of uranium mining itself, onto issues of Native American ethnic and national identity within a history of white conquest and Indian resistance. Reservation policeman Walter Crow Horse thus tells how the white boarding school teachers forced him to give up his language by washing his mouth out with water. The specifice environmental issue, when finally revealed, thus becomes another example of a history of oppression.

The movie represents elements of traditional Sioux culture, such as Pow Wows and shape-shifting, in a picturesque but relatively nuanced way. For example, the tribal elder communes with owls and the wind, but also watches Mr Magoo on television: the modern and the traditional co-existing in a way that avoids the sentimental images of Native Americans as noble savages perpetuated by movies such as *On Deadly Ground* and *Dances With Wolves*.

Moreover, *Thunderheart* also constructs the Sioux as agents of resistance and advocates of environmental justice. Dartmouth-educated Native American rights activist Maggie discovers that her people are getting sick because their water supply is contaminated. When warned by Ray not to get involved, she snaps back: 'My family have been involved since Columbus'. For Maggie, 'power is that river, right there. And that's what I have to protect, not the white law'. Fellow activist Jimmy Looks Twice voices the reasons for fighting for environmental justice in terms of political self-determination and ethnic identity. 'We choose the right to be who we are', he says. 'There is a way to live with earth, and a way not to live with earth. We choose to live with earth'.

In contrast, FBI agent Frank Coutelle voices the arguments of the colonizer. They are a proud people, he tells Ray, but a conquered one, 'and that means that their future is dictated by the nation that conquered them. Now rightly or wrongly, that's the way it works, down through history'. Ray calls this, bitterly, 'protecting the integrity of the American dream'. Yet Ray is radicalized by what he sees on the reservation, and becomes a renegade hero, reclaiming his Sioux heritage in order to protect his people, by identifying himself as Thunderheart, his ancestor at Wounded Knee.

At the end of this revisionist thriller, there is a car chase, but no shoot-out. Ray asserts Indian interests against those of the white developers, telling them: 'This land is not for sale'. The Sioux win a temporary victory, but the ending is open and unresolved. Ray has learned that knowledge is a key to collective power. 'The people know', he tells Coutelle. 'You can't kill all of us'. Ray understands that there will be an Internal Security investigation—that is, a whitewash. But he will use Maggie's media contacts to 'see what kind of a story we can tell'. Ray has discovered a group identity, and *Thunder-*

heart ends with a sense of collective resistance unusual for a genre usually obsessed with celebrating individualist action. The movie also differs from those discussed in this chapter in its relatively unsentimental representation of Native American culture, and its willingness to engage with the Sioux as both a contemporary people and one involved in issues of environmental justice. The existence of *Thunderheart*, then, demonstrates that there is nothing inevitable about the misrepresentation of Native American cultures in Hollywood cinema.[18]

FIVE

The Politics of the Amazonian Rain Forests

Hollywood's interest in the rain forests as an explicitly ecological space is an effect of wider historical developments. In the late 1980s scientists revived theories of global warming first formulated as far back as the 1890s, citing the carbon dioxide produced by the large-scale burning of the world's rain forests as an important cause of the 'greenhouse effect'. By 1992, global warming had become the subject of the United Nations summit in Rio. Andrew Ross comments that First World attention to the ecology and politics of those Third World countries in which the rain forests are located only began in earnest when deforestation was seen as a threat to global ecology, and therefore to the interests of the First World itself.[1] First World media attention to the rain forests was further reinforced with the murder in December 1988 of Chico Mendes, President of Brazil's National Council of Rubbertappers, by ranching interests opposed to his plan to create an extractive reserve in the forest that would be off-limits to logging.

An indication of Hollywood interest in rain forest preservation came in February 1990, when the recently formed Environmental Media Association held 'An Evening For Brazil' in Los Angeles, a benefit for the Rainforest Foundation, founded in 1989 by Sting, Trudie Styler and the Menkragnoti Kayapo Indian Chief, Raoni. Yet the Hollywood movie companies approached the idea of making movies about the rain forest with typical caution. The movies that were produced in the early 1990s tended to evade the social and political issues that lie at the centre of the ecological problems of the rain forests. As such, they tend to confirm the argument made by Susanna Hecht and Alexander Cockburn that First World interest in the rain forest and its peoples has been a mixture of genuine concern and opportunism that 'attains the thunder and velocity of an avalanche but, alas, an avalanche that drops into the void between solicitude and political reality'.[2]

Hollywood's caution was shown clearly in the summer of 1990, with the scramble for the rights to the Chico Mendes story, particularly focused on Andrew Revkin's book *The Burning Season.* Although David Puttnam finally won the rights for Warners, who apparently pencilled in Robert de Niro, Dustin Hoffman or Andy Garcia to play the lead role, the project was put on hold, a victim, according to Puttnam, of Hollywood's reluctance to take risks with political subject matter.[3]

The movies that *were* made by Hollywood in the early 1990s drew on a range of genres, including comedy, action-adventure and the musical. The following examination of the ecological and political assumptions in Hollywood movies about the rain forest will centre on the reasons they posit for the threat to the rain forests, and the type of solutions, if any, that they propose or imply. It will also consider the ways in which these fictions dramatize the encounter between Anglo-Americans and the indigenous peoples of the forest. Hollywood's construction of the rain forests tends to recapitulate the myths of pristine wilderness and the ecological Indian already examined in this book. In this neo-romantic, arcadian fantasy, nature is a pristine Garden of Eden, and Native Americans are a part of that nature, living in innocent, non-destructive, unchanging harmony with their natural environment. In addition, the deep ecological Indian is a figure of natural man against which the hegemony of the white American male can be renewed.

In her essay 'Amazonia as Edenic Narrative', Candace Slater usefully distinguishes between three key social constructions of nature: wilderness, jungle and rain forest. 'Although travelers may lose their way in either a wilderness or a jungle', she writes, 'the latter is distinguished less by the vast solitude that makes it a fitting stage for either contemplation or heroic action than by its disordered and disorienting growth'.[4] The term 'rain forest', on the other hand, derived from botany and plant biology, only entering the general vocabulary in the late 1970s, coinciding with the rise of both environmentalism and state-sponsored economic development in such regions. Accordingly, the term brings with it different connotations from that of 'wilderness' or 'jungle'. A rain forest, observes Slater, 'may sustain different sorts of settlement and cultivation without any threat to its identity', and therefore implies a greater openness to human uses than is possible in a wilderness. 'Rain forest' also implies vulnerability, and, in contrast to the desire to leave wildernesses alone, a 'perceived need for active intervention' in order to save it from destruction.[5]

Film critic Erik Davis observes how the cycle of ecologically themed rain forest movies in the early 1990s were revisions of the earlier trope of the 'jungle'. In jungle narratives, a white European male, in contact with a natural world experienced as threateningly demonic and Other, typically descends into atavistic madness or violence, or else succumbs to a total immersion in the Other; in other words, he 'goes native'. This is the mythic area explored by the German film-maker Werner Herzog in *Aguirre: The Wrath of God* (1972) and *Fitzcarraldo* (1982), and in American cinema most notably by Francis Ford Coppola's *Apocalypse Now* (1979). Similarly, in *The Mosquito Coast* (1986), the jungle became a scene in which the hubristic ambitions of male Westerners re-enact and perpetuate imperialist ventures. As Davis writes, 'For the fathers, the jungle is where you perform the ultimate identification with the Father: to look upon the oceanic maternal chaos . . . and erect an image of your will'.

The recent cycle of explicitly 'green' movies, Davis observes, were an attempt to revise the jungle narrative, particularly in relation to the forests' indigenous inhabitants. 'Historically', he writes, 'Westerners have found that the best way to deal with the ambiguity of the jungle alien is to kill it, enslave it, convert it, or study it. Distancing themselves from the decidedly mixed legacy of the Western, green movies attempt to "ethnologize" the savages into peoples just as they "ecologize" the jungle into the rain forest'. Davis also makes the point that the rain forest has become a convenient visual icon for an ecology movement whose concepts are often invisible and abstract.[6]

Crucially, these mythic projections onto the land and its peoples have a political dimension. The history of First World tutelage of the Amazon, in which governments, anthropologists and so-called development experts speak for the Indian tribes, amounts to what Hecht and Cockburn call 'a refusal to permit the Amazon to tell its own story'.[7] In cinema, this attitude has informed ethnographic documentary films, especially in their reliance on voice-overs that purported or assumed to give a 'scientific' account of native peoples, delivering the 'truth' about them while denying them a voice in reply. This elitist model has been questioned in more recent documentary work by various aesthetic strategies centring on more participatory or collaborative modes of production, which, as Shohat and Stam argue, 'discard the covert elitism of the pedagogical and ethnographic model in favor of an acquiescence in the relative, the plural, and the contingent, as artists experience a salutary self-doubt about their own capacity to speak "for" the other'. [8]

Unsurprisingly, given its commercial imperative, Hollywood cinema has responded slowly to such critiques of established representational practices. Instead, it has continued to rely on old narrative formulae, particularly those in which indigenous peoples and their environment become picturesque or exotic backdrops to assertions of white, First World heroism. Moreover, as Erik Davis points out, the indigenous tribes represented in these movies tend to be the fictional constructions of Indian choreographers and language creators. Authenticity and truth-value are signified by techniques appropriated from ethnographic documentary, such as the use of subtitling and long, uninterrupted takes. However, as Davis notes, 'the ethnographic impulse generally fails to deliver either cultural depth or force of character. . . . Because these movies are uncomfortable with Indian agency, even natives with lots of lines seem like extras, their conversation heavy with mysticism and short on information or plain humanity'.[9]

John Boorman's British-produced movie *The Emerald Forest* (1985) began the cycle of explicitly ecological rain forest movies. As in the director's earlier film *Deliverance*, discussed in Chapter Three, the threat to the pristine wilderness comes from the construction of a dam, an emblem of modern technological development. Moreover, the natural environment again becomes a testing ground for traditional forms of masculine heroism.

Perhaps the most contrived connection with rain forest preservation came in *The Forbidden Dance* (1990), released to cash in on the dance craze in the title. The movie begins solemnly with the title: 'Brazil—The Amazon: Mankind is destroying the rain forest'. The opening title is, however, the movie's last attempt at authenticity. In the opening sequence, picturesque and beautiful tribespeople dance the lambada, when a Texan oil boss arrives in a jeep and gives them five minutes to leave the rain forest. The tribal princess, Nisa, played by former Miss USA Laura Herring, and a bald witch doctor then go to Los Angeles to appeal on national television to save her 'ancestral home' from destruction. Nisa equates the lambada with closeness to nature, teaching her white American friend Jason to dance with the words: 'become a tree with deep roots, flowing with life, and everything else will follow'.

Nisa then tells Jason about the destruction of the rain forest. 'They must stop killing the trees', she tells him, 'or the sun will eat the air'. Somewhat patronizingly, Jason develops this idea for her, in a speech laughable in its muddled and inaccurate pseudo-scientific rhetoric: 'You're talking about the hole in the ozone layer. And this hole is getting bigger in part because of the burning of the rain forest'. Jason goes on to confirm his commitment to Nisa's cause: 'This is bigger than you know. This is not only important to you people. This is important to people everywhere. And it's important to me'. The couple eventually win an audition to dance the lambada on television, so that Nisa can deliver her ecological message to the nation. Yet the movie again gives the authoritative word to a male, rather than to the indigenous Indian woman, as the singer Kid Creole calls on the television audience to boycott the oil company's products in the supermarkets (although what the company actually makes is not specified). The movie ends with a title dedicating it 'to the preservation of the rain forest'. *The Forbidden Dance*, then, attempts to reconcile ecological awareness with consumerist fashion, to create a movie that is both vacuous and risible in equal measure.

The following year saw the release of *Meet the Applegates* (1990), which uses the issue of rain forest depletion as the basis for a comedy-horror film satirizing the hypocrisy and moral corruption of small-town America. At the start of film, the Edenic rain forest is being cleared by cattle ranching interests: a sign indicates a Beef Products Manufacturing Facility funded by the National Bank. The narrative then shifts to a family of insects which disguises itself as average Americans while plotting to sabotage a nuclear power station, in order to wreak revenge on the human beings responsible for destroying its forest home. The movie ends, however, with an endorsement of 'non-violent civil disobedience' as a more ethical and effective way of protesting against the destruction of the forests. Resistance is, however, limited to the insects and a small group of American sympathizers, with the Indian inhabitants of the forest sidelined as comic caricatures with painted faces.

At Play in the Fields of the Lord (1991), in marked contrast, is a more realist and complex treatment of the Amazonian rain forest and its peoples. The rights to Peter Matthiessen's novel were acquired by MGM in 1966, the year after its publication. The project subsequently went through several re-writes and changes of personnel in a period of over twenty years before finally being given to Argentinean-born Hector Babenco to direct.

Babenco decided to use citified Indians in acting roles, rather than involve tribes indigenous to the forests, thereby showing his unwillingness to reproduce the neo-colonialist exploitation of indigenous peoples apparently practised by Herzog and Coppola when making their movies.[10] In doing so, he rejected his producer's idea to film actual Yamomani Indians on their own land, telling the press:

> The Indians would love to have the movie there. They knew it would mean gifts, things to trade, maybe they could buy boats. But if we did that, we would be doing the same thing as these [expletive] missionaries are doing to the Indians. I would be trying to explain our concepts of time, obligations, hierarchy . . . Not me. I am not going to use people and throw them away.[11]

Babenco also stated his intention of revising the colonialist narrative strategy according to which the experiences of First World protagonists are explored with a Third World country used merely as an exotic backdrop, asserting that 'The Indians are the most important part of the movie. The Indians are not an excuse for the Anglo-Saxons to do their thing'. [12]

The narrative centres on the encounter between white Anglo-Americans and indigenous Indians, worked out in particular in terms of Judaeo-Christian notions of guilt and redemption. Like the British film *The Mission* (1986), *At Play in the Fields of the Lord* is concerned with Christian contact with indigenous Indians in Amazonia, but is far more critical of its contemporary missionaries than the British movie is of its eighteenth-century Jesuit priests.

Set in Peru, near the Brazilian border, the movie locates the threat to the indigenous forest peoples in gold prospectors supported by the state, as represented by the crafty bureaucrat Commandant Guzman, who is himself in league with the Christian missionary leader. Commandant Guzman wants to remove the Niaruna (the tribe invented for the movie) from their lands to allow the forest to be developed for mineral extraction. The movie develops Guzman's motives and attitudes in a relatively complex way. At one point, he cites poverty as the motive force behind the state's policy of aiding the economic development of the forest. 'My job is to stop these peoples from taking the Indians' lands', he explains, 'but sometimes when people are starving, nobody can stop them'. But Guzman also exhibits a racist, colonialist mixture of envy and contempt for a people whose way of life is different from his own. The 'savages', he says, 'don't need this land. It's too much for a few thousand people who do nothing from morning to night'. Later he tells the Christian missionaries: 'The Amazon is not the Garden of Eden. The people here, they want progress. They must face the reality now or later'. Yet his euphemistic language disguises the threat of violence which he eventually exercises against the indigenous Indians, when at the end of the movie he orders a helicopter to bomb the Niaruna village into submission, in order to evict the tribespeople from their lands. The leader of the Protestant Christian mission, Leslie Huben (John Lithgow), had earlier admitted to his wife Andy (Darryl Hannah) that he did a deal with Guzman. 'He gave us a year to pacify the Niaruna', he tells her, 'and then he would do the job in his own way'.

At Play in the Fields of the Lord avoids the patronizing reduction of the Amazonian tribespeople to child-like innocents, evident in *The Emerald Forest* and *The Mission*, while constructing a complex sense of their ecological relationship with the forest. The disillusioned Catholic priest Father Xantes, who has come to regret his mission, mourns what he refers to as the Indians' 'misfortune' in being converted to Christianity. Xantes has come to learn the connection between environment, religion and cultural imperialism. 'Allah, Christ will never be accepted here', he tells Andy. 'A pale man from a desert country where it never rains'. Accordingly, the padre voices the most respectful view of Indian culture in the movie: 'Who knows what we might learn from our poor Indians, if we're not always teaching them. Such an easy people, light as the air, like a leaf or a cloud. They do not seek for meaning the way we do, they just are'. Significantly, director Hector Babenco reiterated a similar argument in a press interview: 'These human beings live in perfect harmony with their environment; they are the real keepers of the forest . . . Then we white people tell them how great our civilization is, how great our God is. The transparent universe is broken'.[13]

The movie also avoids the tendency in Hollywood cinema to homogenize Native Americans. Instead, characters are sufficiently differentiated to allow for differences of opinion to emerge within the tribe over how to deal with the threat of white invasion. Moreover, though sequences depicting religious practices, such as shamanic trances, are filmed at length, displaying Indian culture as anthropological spectacle, they have an intensity that avoids the quaintly picturesque. These sequences also suggest that Native religion is more adaptive to the local environment than the Christianity practised by the missionaries, which is represented variously as paranoid, sexually repressed, guilty, and either dogmatic or doubt-ridden. In contrast, the Indians embody unrepressed desire, their nakedness signifying a state of spontaneous nature.

Significantly, and unusually for an American movie dealing with white encounters with the wilderness, *At Play in the Fields of the Lord* denies its Anglo-American characters the possibility of redemption through contact with pristine nature. In fact, the movie represents contact between the Indians and the white Europeans as an unmitigated disaster for both races. This issue is particularly focused in the character of Lewis Moon (Tom Berenger), the half-Cheyenne soldier-of-fortune who regrets that, during the nineteenth-century Plains Indian wars, the Cheyenne, Crow and Shoshone had resorted to killing each other, 'because we were just dumb fucking Indians too fucking stupid to recognise our real enemy'. Moon tells the liberal missionary Martin (Aidan Quinn) that if the latter genuinely respects Indian language and customs, he will not try to change them. 'If the Lord made Indians the way they are', he comments, 'who are you people to make them different?' Hired by Commandant Guzman as a mercenary, Moon refuses at the last moment to bomb the Niaruna village. Instead, he tries to go native: tripping on the local hallucinogen, he strips naked, and attempts to rediscover his repressed Indian self. Having fallen from the sky (in a parachute), he is taken by the Niaruna for their sky god, Kisu the Thunder Spirit. Yet, unlike the father and son in *The Emerald Forest*, or John Dunbar in *Dances With Wolves*, Moon's attempt to 'go native' is unsuccessful, and he is finally unmasked as a false god. Having warned the Niaruna that it is in their interests to avoid all contact with the whites, Moon foolishly acts contrary to his own advice, when he kisses Andy Huber. This erotic contact is disastrous: he catches her cold, which he then passes on to the tribe, with fatal consequences. As Martin says to Moon: 'if you'd stayed away from the mission, this epidemic would never have occurred'. Moon is responsible for weakening the Indian tribe, his attempt to go native proving to be merely an act of foolish, self-indulgent play-acting.

In *At Play in the Fields of the Lord*, then, inter-racial contact is disastrous. The white missionary child Billy strips naked and plays happily with the Indian children, but dies of blackwater fever carried in a mosquito bite. Bereaved, paranoid and alienated from her 'big, ugly body', his mother Hazel (Kathy Bates) goes mad, and has to return to the United States. Her husband Martin, the liberal, compassionate Christian who respects Indian culture, begins to doubt the methods used by the Protestant mission effectively to bribe the Niaruna into becoming Christians, and finally reaches the conclusion that 'it would have been better for them never to have known us'. Yet there is no deliverance or redemption for Martin either: he is killed, ironically, by his Christianized Indian guide, as the government helicopter begins to bomb the Niaruna village. The film ends with a slow, aerial tracking shot away from Martin's body to reveal the burning rain forest as a stark fact with no solution offered.

Medicine Man (1992) also deals with the fragility of Indian culture, and problems of contact and contamination by white invaders. The threat to the rain forest in this movie comes from a logging company, which is clearing the forest to build a road, and is supported by the government's plans to relocate the tribal peoples. Typically for Hollywood movies about the rain forest, however, the point-of-view of those deemed responsible for destroying the forest is not entered into. Instead, the loggers are nameless and violent, and their technology is demonized: the bulldozers are filmed at night, in an infernal scene of white floodlights, expressionist camera angles, and cacophonous noise and drumming on the soundtrack. The burning of the forest is equally apocalyptic, with backlit figures running in slow motion, again accompanied by ominous music.

Ultimately, *Medicine Man* implies that the interests of the loggers should be overridden by an alliance between the tribal peoples and Western capitalist medicine. The narrative concerns the successful overcoming of obstacles to this process of cultural reconciliation, particularly centred on the troubled subjectivity of the white European male, Dr Robert Campbell (Sean Connery).

Campbell is wracked with guilt, believing that when he worked for an American drug company looking for quick profits in medicinal forest produce, he was responsible for bringing swine fever to an Indian village. The local shaman believes that the disease was the gods' punishment for the indiscretion of sharing his 'forbidden juju magic' with the whites. The shaman's distrust of Campbell is compounded when the latter cures a sick child with Alka-Seltzer: the shaman takes off, because, as Campbell puts it, 'I had taken his stick . . . his self-respect. My intentions were well meant'. In retaliation, the tribal medicine man refuses to tell Campbell where the medicinal plant he is looking for is located.

In the end, Campbell overcomes his guilt, and when his assistant Rae (Lorraine Bracco) administers a hypodermic injection to save the life of a sick native boy, the shaman forgives Campbell. The whites, through their scientific know-how, thus become the saviours and protectors of the forest Indians. Moreover, their subsequent discovery of an anti-cancer serum is a triumph for high-tech medical science, the emblem of which is the computer they set up in the forest. In the event, Rae makes the vital scientific breakthrough with the help of serendipity, when she discovers that the source of the serum is not in the plants they are testing, but the ants that have contaminated the sugar used in their experiments. In this way, capitalist technoscience is renewed, and the American pharmaceutical firm grants the maverick (pony-tailed) Campbell new equipment and funding to continue his research. In terms of ecological politics, *Medicine Man* thus reflects a First World conservationism that seeks to make the rain forest what Hecht and Cockburn call 'an Eden under glass', using the Amazon, as they put it, as 'an enormous, unsullied laboratory for the scientific contemplation and classification of nature'.[14] *Medicine Man* posits this relationship as a reciprocal partnership between native Indians and European-Americans, who can harmonize their different interests in conserving the rain forest for their mutual benefit. The forest will provide medicinal compounds for big capitalist pharmaceutical companies, while its tribal peoples can continue to live in their traditional way in their Garden of Eden.[15]

However, this supposedly egalitarian outcome to the movie is problematic, in that the political issue of whether such a strategy will be an act of capitalist appropriation of tribal knowledge, or a genuine partnership with tribal peoples, is not seriously addressed. Instead, the movie recalls *The Emerald Forest* in its patronizing idealization of its native Indian tribe, the fictional Poca Nu, who are depicted, like the Invisible People in Boorman's movie, as a child-like, joyful people, near-naked and playful, with the camera tending to linger voyeuristically on naked female bodies in scenes accompanied by jaunty panpipe music on the soundtrack. The main role played by the Poca Nu in *Medicine Man*, then, is that of purifiers of the fallen European soul, as both white protagonists are redeemed through contact with primitive, wild nature. Rae, a cynical woman from the Bronx, learns humility and a deeper knowledge of life through contact with the forest peoples and through her growing love for Campbell. 'Life is strange', she muses at the end, 'but down here it seems . . . so very precious'. Rae successfully goes native (the blue tattoo that the Indians paint on her face is indelible), and is rewarded with success both in love and in her career, gaining at the end of the film both the love of Dr Campbell and 'joint publication'.

In parallel, Campbell's authority and self-respect are restored through the shaman's message of forgiveness and respect, as relayed to him by Rae: 'He said he'll teach you big magic. He says he's never met a man like you before, and maybe he never will again'. The relationship suggested here, then, is more one of Native American deference for the white male hero than of an equal partnership. The wilderness has again become a proving ground for the white European male, who has demonstrated his prowess both to the native people of colour and to his female companion. At one point in the story, Campbell even gets to act like Tarzan, swinging from the vines to save a stoned and careless Rae. White male hegemony is thus again restored in the pristine rain forest of the movie.

In *The Fate of the Forest*, Hecht and Cockburn argue that rain forest depletion is being brought about by the 'unmanaged, clumsy and brutal' use of fire for land clearance purposes by ranchers, large-scale landowners and landgrabbers, who are responsible for 90 per cent of the land that is deforested. The authors locate the causes of this environmental degradation not in a demonized industrial technology, nor, perhaps more surprisingly, in international capital, but in 'a philosophy and strategy for regional development formulated by the Brazilian military' since it seized power in 1964.[16] Policies of land reform and modernization have led to the 'development' of the forests for ranching, mining and logging. Of these factors, the expansion of pasture land for cattle ranching is the main reason for deforestation. Yet, contrary to popular misconceptions, such as to be found in the Hollywood comedy *Meet the Applegates*, Hecht and Cockburn argue that ranching in the Amazon has 'nothing to do with North American fast food'. In fact, the Amazon is a net *importer* of beef. Instead, cattle are used 'primarily as an excuse for claiming land, for clearing it and for economic purposes that have little to do with producing commodities'. In other words, in an economy where cleared land is more valuable than forested land, land speculation is the most important factor in deforestation, especially as a hedge against spiraling inflation.[17] Tom Athanasiou similarly relates the problems of deforestation and the lack of sustainable development in the Amazon to issues of land ownership, and the political need for land reform. In Brazil, he observes, 0.8 per cent of the landowners own 43 per cent of the land. Landless peasants migrate into the forests to escape poverty, while governments promote rain forest colonization as a useful safety valve to mitigate potential social discontent.[18]

Of the movies discussed in this chapter, only *At Play in the Fields of the Lord* attempts a complex and nuanced dramatization of the issues involved in the deforestation of the Amazon. The other movies tend to find recourse in the simpler strategy of separating the ecology of the rain forest from issues of social justice. Again with the exception of *At Play in the Fields of the Lord*, Hollywood's discursive construction of the Amazon rain forest has also tended to perpetuate the notion of the region as a pristine Garden of Eden, a notion that ultimately reinforces a preservationist politics that would effectively turn the Amazon into a national park, in the words of Hecht and Cockburn, 'secluded by law, force, and cash bribe (the debt-for-nature swap) from the predations of man'.[19] The authors argue that the emergence of what they call an 'ecology of justice' in the Amazon has been hampered by the way in which tribal peoples are still perceived by Europeans in terms of myths of an Edenic nature. Yet, as argued in Chapter One of this book, the supposedly 'primal' American 'wilderness' is not static and unchanging, but is part of a history of human intervention which includes the slash-and-burn agriculture practised for centuries by tribal peoples themselves. As Andrew Ross puts it, it is

'Western', urban anxieties about modernity that construct a compensatory neo-romantic myth of peoples untainted by commercial contacts. This condescending fantasy neglects the fact that the tribal peoples in Amazonia have a complex history that interacts with that of the small extractors (mainly of gold, nuts and rubber), the World Bank, national governments and multinational capital.[20] The conflict over resources and way of life is therefore a complex issue, which Hollywood's framing of its melodramatic narratives as Manichean moral conflicts does little to address, both in their tendency to construct Indians as simple innocents and their parallel refusal to attribute a complex perspective to the figures they posit as the enemy, be they the cattle ranchers in *Meet the Applegates*, the loggers in *Medicine Man*, or the dam-builders in *The Emerald Forest*. Only *At Play in the Fields of the Lord* makes an attempt to account for the actions of its gold prospectors in terms of the poverty that is a crucial factor in calls for the development of the Amazonian forests. Thus, in concentrating on visual icons of ecological destruction, such as logging machines and forest fires, movie-makers may have discovered powerful, instant visual images for environmental disaster, but they have done so in a way that tends to confuse effects and symptoms for causes. The Hollywood movies discussed in this chapter therefore tend to confirm Candace Slater's argument that what she calls the 'luminous distance' of the rain forest in the popular imagination 'encourages the outsider to imagine a struggle between obvious victims (the "virgin land") and equally obvious villains (miners, loggers, ranchers) in which he or she is in no way complicit'.[21]

II

WILD ANIMALS IN HOLLYWOOD CINEMA

Introduction

From the silent era of Hollywood cinema to the 1960s, wild animals tended to be represented in imperial narratives celebrating the conquest of nature by heroic, white European or American males. The wild animal was an obstacle to this narrative of progress, and was accordingly demonized as excessively savage and monstrous. Such constructions of the wild animal have continued to the present day, as the savage silverback gorillas in *Congo* (1995) demonstrate. By the late 1950s, however, a small number of movies began to react to rising popular interest in the conservation of wild animals, who were rehabilitated from varmints to be eradicated to valued, benevolent members of nature reconceptualized as an organic, interdependent whole. Such develop-ments reflected both the popularization of scientific ecology in the mid-twentieth century, and the growing scarcity of wild animals due to habitat destruction. The 'civilizing process which imperils wild nature', writes Roderick Nash, 'is precisely that which creates the need for it'.[1] Yet the permitted 'wildness' of wild animals in the popular culture of the 1950s was heavily circumscribed, as David Peterson del Mar's study of the representation of animals in *Reader's Digest* magazines of the period shows. Animals, he observes, were not approached on their own terms, as 'sentient beings with their own agendas and rights. Rather, they elicited affectionate interest from humans largely on the basis of their purported interest in humans'.[2] Even wild animals were viewed as inherently friendly, and ultimately subservient, to human beings, to the point where they were seen actively to seek human contact in preference to life in the wild.

These attitudes to wild animals may be traced back in North American culture to the humanitarianism of nature writers such as William J. Long and Ernest Thompson Seton. Long, in particular, is an important precursor to the attitudes to wild animals increasingly prevalent in Hollywood cinema from the 1950s on, in his rejection of the morally disturbing implications of Darwinism. As Robert Bannister argues, Social Darwinism (the political ideology which asserted that *laissez-faire* economics and war are both an inevitable part of the extension to human societies of Herbert Spencer's doctrine of nature as the 'survival of the fittest') was largely resisted by the lay public, including businessmen, in the late nineteenth century. Although conservatives often evoked Darwinism to give a spurious form of scientific legitimization to socially produced problems of racial inequality and oppression, in issues other than race, Bannister argues, 'Christian tradition, democratic values, and faith in the harmonies of nature set limits on scientific defenses of power and privilege, at the same time suggesting a strategy (the label social Darwinism) for discrediting all such tendencies'.[3] Even capitalist businessmen resisted the Malthusian-Darwinian-Spenserian model of social relations, because, as Bannister puts it, 'individuals who desire stability, consensus, homogeneity, and peaceful change under a capitalist regime—as did businessmen and many of their middle-class defenders—found little comfort in a cosmology that posited permanent struggle as the engine of progress'.[4] In a similar way, the anthropomorphism of William Long's writings rendered nature acceptable within the limits of bourgeois gentility and Protestant piety, reassuringly muting the struggle for existence by emphasizing animals as altruistic, co-operative and essentially spiritual beings. Nature was therefore constructed to provide lessons in Christian virtues, in a universe that was moral and purposeful.

Nature writers such as Long and Seton helped to extend animal welfare concerns to wild animals, a process reinforced by their tendency to compose stories from the imaginary point-of-view of individual animals themselves, rendered sympathetic and human-like in their psychological motivations and emotions. As Ralph Lutts comments, animals were represented as 'furry or feathery little people', so that readers might care for wild animals 'as they would care for other people, or at least as they would for their own pets'. Nevertheless, he adds, such writing also perpetuated a type of wishful thinking that fostered 'anthropomorphic, sentimental, death-denying, and distorted or partial understandings of the natural world'.[5] Moreover, it used the authority of scientific discourse to reinforce its truth claims, so that, as Robert Elman puts it, 'the romantic humanizing of animals was presented as fact proven by scientific inquiry'.[6] As Part Two of this book will demonstrate, the construction of wild animals by those writers dismissed by Theodore Roosevelt as 'nature-fakers' anticipated attitudes that are still prevalent in American popular cinema today.

Ted Benton summarizes the symbolic connotations attributed to the 'wildness' of the wild animal in American culture: 'a liberty of action and expression, a carefree and dangerous libidinous abandon, unrestrained by the burdens and disciplines of civilized existence, or unaffected by the degeneracy, enfeeblement and dependency of domestication and regulated existence'.[7] In Hollywood cinema, this association of wild animals with a sense of anarchic freedom has increasingly become desirable, rather than dangerous. Indeed, by the 1960s, a new narrative had emerged in popular cinema, one that turned on the release of a wild animal from captivity back into the wild, and thereby celebrated wildness as a state separate from human society in which wild animals had a right to live, and about which human beings themselves could fantasize. The Anglo-American production *Born Free* (1966), which will be discussed at length in Chapter Nine, became a model for later movies of this type, such as the *Free Willy* cycle of the 1990s. Nevertheless, these narratives continue to circumscribe their notions of 'wildness' within limits acceptable to bourgeois society. As Ralph Lutts puts it, a conflict between 'wishing to turn wild animals into cute pets and the desire to ensure their survival as autonomous beings in the wild pervades American society'.[8] Moreover, older forms of theriophobia, in which wild animals remains figures onto which social and psychic anxieties are projected, still remain as tendencies within Hollywood cinema. Indeed, the success of the *Jaws* cycle and its imitators in the 1970s suggest that the need to displace and externalize guilt and fear onto wild animals still exists, despite the rise of conservationism and concern for endangered species. As already mentioned, the gorillas in *Congo*, as well as the monstrous lions in *The Ghost and the Darkness* (1996), continued this trend into the 1990s.[9]

Explicitly conservationist narratives, in contrast, construct wild animals at the opposite end of the Manichean polarity, as benign exemplars of virtuous behaviour. In doing so, they continue the symbolic appropriation of animals for lessons in human social relations, seeking to provide justifications in nature for ideological beliefs, particularly concerning gender and family relationships. As Part Two will show, then, several potentially contradictory representations of wild animals, derived from popular American nature writing, the animal fable, the circus and the zoo, as well as a newer conservationist sensibility, are at work in Hollywood 'creature features'.

Part Two of *Green Screen* attempts to examine the anthropomorphic attitudes projected onto wild animals in Hollywood movies by taking account of the real animal external to such discourses. The critical realist position assumed here thus runs counter to poststructuralist philosophies of nature which have tended to deny, or at least obscure, the existence of a non-human nature external to human discourses about it. As mentioned in the Preface, the main theoretical basis for the position taken in this book is Kate Soper's *What is Nature?*, which argues for a philosophical position between naïve realism on the one hand and extreme social constructionism on the other.

An example of the problems created by an extreme social constructionist view of animals comes in the study of zoos by Bob Mullan and Garry Marvin. The writers assert that 'once the animal becomes the focus of human attention, the notion of the "real" animal makes no sense, for all animals as perceived by humans are the result of human interpretations'.[10] The fallacy here is the implication that animals are *only* the result of human interpretations, an assumption which does not take into account biological and ecological processes in which human beings may not play a significant part. Wild animals, on the contrary, have an existence independent of human perceptions of them. The critical realist approach adopted in the following chapters is therefore against the relativism implied by Mullan and Marvin's approach. Instead, it assumes that some social constructions of animal ethology are more accurate than others, when tested empirically against other discourses on the real animal.

An approach of this kind allows for a challenge to the anthropocentric assumptions of much film criticism, of which the critical reception of *Jaws* (1975) is a good example. For most critics of the movie, the great white shark is simply a metaphor or symbol. Michael Ryan and Douglas Kellner, for example, assert that the shark is 'the sign of what goes wrong when the male sex is not fulfilling its duty of patriarchal leadership'.[11] Stephen Heath implies that the choice of great white shark as antagonist is arbitrary: 'the evil is something else ...', he writes, 'call it a shark'.[12] While these writers produce perceptive interpretations of the symbolic connotations of the wild animal, they tend to displace critical attention away from the representation of the animal itself. In response to such criticism, the following four chapters provide an environmentalist critique of the representation of wild animals in Hollywood movies, by exploring how the symbolic meanings attributed to them interact with literal, but nevertheless provisional, constructions of the real animal. In assuming that there is a 'true' or 'real' animal, against which a given representation can be measured and found to be a distortion or misrepresentation, the critical strategy employed here draws on realist debates on the politics of representation developed for human groups subordinated in terms of gender and race, for example. Even though they are prone to simplistic and crude applications, such inquiries into stereotyping and misrepresentation become politically important when such representations are understood as contributing to unjust social effects. Although the possible effects of the movies discussed in the following pages on the treatment of actual wild animals is beyond the scope of this study, it is nevertheless important for an environmentalist criticism to explore the representation of wild animals in cinema in a way that attempts to decentre the overly anthropocentric approach of much film criticism.

SIX

North American Anti-Hunting Narratives

The figure of the hunter has been central to American notions of white masculine identity, inscribing patriarchal gender relations within traditional wilderness mythology. Hunting was promoted by Theodore Roosevelt, who co-founded the Boone and Crockett Club in 1887, for its role in inculcating the militaristic leadership qualities necessary for the white Anglo-Saxon managerial class to triumph in the Social Darwinist war of nature.[1] Hollywood cinema has subsequently celebrated the activities of white hunters and trappers by casting charismatic actors who embodied the rugged individualistic values of traditional American heroism. In *The Adventures of Robin Hood* (1938), starring Errol Flynn, the rewards of hunting were given as part of nature's abundance, providing the band of noble outlaws with food. The western *How the West Was Won* (1963) cast James Stewart and Henry Fonda as trappers, displaying thereby a nostalgia for the individual freedom and closeness to nature associated with the traditional wilderness hero.[2]

Nevertheless, a small body of films have espoused anti-hunting sentiments. *Bambi* (1942), discussed in Chapter One, was followed by a cycle of explicitly anti-hunting movies in the 1950s, at a time when public perceptions of wild animals were changing. In more recent times, the post-Vietnam era produced *The Deer Hunter* (1978), in which the Natty Bumpo figure Mike (Robert de Niro) can no longer bring himself to kill the deer he has in his gunsights, his experience of killing and being hunted in Vietnam giving him a new sympathy for his former prey. In contrast, John Milius's right-wing survivalist movie *Red Dawn* (1984) included the ritual drinking of deer's blood by white vigilantes, as hunting was again seen to guarantee manly strength in the Darwinian war of nature.

Compared to the hunting practices of white hunters, those of Native Americans have continued to be represented in a wholly positive light, as movies draw on the notion of the ecological Indian discussed in the Chapter Four. At the start of *The Last of the Mohicans* (1992), Native American deer hunters ask forgiveness of the deer they kill: 'We're sorry to kill you, brother. We do honor to your courage and speed, your strength'. The movie represents the deer hunt as part of the idyllic life in the forest threatened by the intervention of the colonial war between the French and the English. Moreover, Native American hunting practices are seen as a harmonious interaction between human beings and nature, spiritually redeemed by religious ritual. 'All animals had souls, of course', writes radical environmentalist Kirkpatrick Sale, 'so in all hunting societies, some form of ritual apology and forgiveness was necessary before the kill'.[3] As this chapter will show, the notion that white hunting is predatory, whereas Native American hunting is both a spiritual act and ecologically benign, is a given in many environmentalist movies.

Buffalo hunting: from Bill Cody to John Dunbar

Buffalo Bill Cody epitomizes for American culture the hunter as master and conqueror of nature through justified violence. Cody boasted in his autobiography of killing 4,280 buffalo in an eighteen-month period, when in 1868 he worked as a buffalo hunter for the Kansas Pacific Railroad, and escorted millionaires and European royalty west to kill buffalo for sport.[4] Cinema has been central to the myth of Buffalo Bill, not least in the films in which he starred as himself, beginning with Edison's *Buffalo Bill and Escort* (1897).[5]

Buffalo Bill (1944) maintained the heroic status of Bill Cody (Joel McCrea) by carefully distancing him from complicity in both the demise of the Plains Indians and the wantonness of the buffalo hunt. Cody is shown expressing regret for the forces of nature paying the price for the inevitable march of American progress, even though he himself is part of that process. This ambivalence is typical of the American wilderness hero, whom Richard Slotkin describes as preferring to live outside of civilization, close to wild nature and to the naturalized values of force and violence it represents, but also as an agent of the commercial development of that wilderness. The hero is, then, a mediator between wilderness and civilization, 'whose experiences, sympathies, and even allegiances are on both sides of the Frontier'.[6]

In keeping with this traditional American myth, the 1944 biopic represents Cody as a friend of the Indian. Nevertheless, although he tells Louisa (Maureen O'Hara) that 'Indians are good people, if you leave them alone', he also does his patriotic duty in fighting the Cheyenne in order to uphold the interests of the white settlers. A similar ambivalence informs Cody's attitude to the buffalo, as the movie sanitizes the full extent of his role as buffalo hunter. The movie begins as it means to go on, as footage of a buffalo herd being chased and fired at by a single horseman, identified by the voice-over as Buffalo Bill, sanitizes the violence of the hunt by filming the action entirely in long-shot, and stopping short of the kill. This evasiveness sets the tone for the later sequences which treat Cody's hunting activities.

Buffalo Bill goes some way to eliciting sympathy for the Indians as victims of American progress. The Cheyenne warrior Yellow Hand believes that the extermination of the buffalo by the white hunters is a deliberate ploy to threaten the survival of Native American tribes dependent on the animal for food and hides. 'The white man has done this thing', he tells a meeting with the Sioux, 'so the red man will starve'. However, the movie is careful not to implicate Bill Cody in such a strategy, nor even in the slaughter of the buffalo for sport. When Cody's father-in-law appoints him head of field operations in his business exporting buffalo hides, he tells him: 'We want all the hides you can get, and more'. This scene is followed by explicit footage of buffalo being shot at, felled, and their hides being carted away. The narrator explains: 'The craze for buffalo robes swept the East. They brought big prices. Buffalo hunting became an organized business, and degenerated into a wholesale slaughter. For a time it became the sport of the world, bringing sportsmen from every land, and in a single month, five thousand head were slaughtered'. A montage of newspaper headlines and advertisements then indicates the arrival in the West of European royalty to take part in the craze for sport shooting from railroad cars. However, in the next scene, Buffalo Bill is shown chasing buffalo in a way that again tempers his complicity in the violence of the hunt: he is not even shown shooting at a buffalo, still less killing one. Instead, the people who *are* shown killing the animals are stereotyped as trophy-hunting, dandified Eastern tourists. It is they, the montage suggests, who are responsible for eradicating the buffalo and starving the Plains Indian tribes, rather than Buffalo Bill.

Moreover, in the following scene at the campground, Cody expresses his concern to Ned Buntline about the large numbers of buffalo being killed. 'This started out as a business proposition', he tells him. 'Now they're shooting them out of train windows for sport. There's a limit to even the buffalo, you know'. Buntline is sceptical: 'Limit? Oh yes, my boy, I suppose there is a limit to the sands of the seashore. But who's going to count them?' Cody replies: 'Maybe you're right, Ned, but it worries me'. He then refuses to meet the Grand Duke, saying he is too tired even for European royalty.

Cody, then, is carefully drawn in these scenes as morally innocent, in that he kills buffalo as a business proposition, but neither for sport, to entertain visiting European royalty, nor to bring about the decline of the Indians. He is also aware of possible limits to their abundance. Through these sleights-of-hand, the movie is able to end with the narrator reaffirming Cody's heroic reputation: 'His name came to typify for all of us frontiers and freedom, adventure and fair play. The spirit of the West'. Jewett and Lawrence argue that acts of denial of this kind are central to the myth of the white redeemer in American popular culture, enabling as they do Manifest Destiny to be presented as a benign process. The 'desire for profitable western lands', they write, 'is ritually repressed by the symbol of the redeemer who seeks nothing for himself'.[7]

The Last Hunt (1956), in comparison to the equivocations of *Buffalo Bill*, explicitly viewed buffalo hunting as wasteful and cruel, and the perverted outcome of a warped and destructive masculinity. The movie is set in 1883, by which time, as the opening title puts it, hunters and Indians had 'recklessly slaughtered' the buffalo herds, reducing their numbers from sixty million to three thousand in thirty years. Sandy McKenzie (Stewart Granger) is a former meat hunter for the Army Engineers. Decent and peaceable, he is weary of killing buffalo, understanding the destruction that over-hunting is causing, not only to the animals themselves, but also to the Plains Indian population. 'It's a crime against nature and the Indian to kill the buffalo', he says. McKenzie plans to settle down and become a farmer, his character thereby enacting Frederick Jackson Turner's notion of American progress from a hunting to an agricultural economy.[8] However, McKenzie's need for money forces him to go on one last buffalo hunt with Charles Gilson (Robert Taylor). Gilson is a troubled protagonist familiar in post-World War Two American cinema: a man trained by war to be psychopathic, violent and racially prejudiced. As McKenzie says of him: 'I've never known a gun to wear a man before'. The following exchange between the two characters makes clear Gilson's belief that violence and killing are essential and unchangeable parts of human nature:

> Gilson: Killing's natural.
> McKenzie: Not to me, it ain't.
> Gilson: Sure it is. War taught me that. The more you kill, the better man you was. Killing, fighting, war, that's the natural state of things. Peace time's only the resting time between, so's you can go on fighting.

Through the figure of Gilson, *The Last Hunt* investigates the self-hating, alienated psychology of a white hunter.[9] The skinner Woodfoot (Lloyd Nolan) suggests that there is an erotic component in Gilson's desire to kill, when he asks him: 'How does it feel to kill so many buffalo? Feeling that you got the almighty power of life and death in your hands? Does it make you feel big, special? Is it the same kind of feeling you get round a woman, maybe?' Gilson replies: 'It's like something important's gonna happen, something that can't be changed back to the way it was ever again. Killing's like the only real proof you're alive'. When Woodfoot warns Gilson that over-hunting has led to a decline in the buffalo population, the latter, like Ned Buntline in the 1944 *Buffalo Bill*, denies that hunting is a threat to the buffalo. 'There ain't no end to the buffalo', he complacently replies.

The Last Hunt links Gilson's desire to kill buffalo with his role as a racial and sexual oppressor. The sexual side of his desire for mastery over the innocent is developed in his relationship with the Indian girl (Debra Paget). She and her baby are the only survivors of an Indian family whom Gilson shot in the back, when poverty had driven them to steal his horse. Gilson claimed ownership of the girl, but she plays dumb, passively resisting his advances, yet staying with him, for reasons she later explains to McKenzie: 'My people are starving. This child must live. So I stay'. McKenzie, in contrast to Gilson, has a proper respect for the Indians. 'I was raised round your people', he says. 'They taught me how to live'. He also understands that Indian cultural attitudes are different from those of white people: 'Stealing horses ain't a crime to an Indian. They don't think the way we do'. But to Gilson, Indians 'ain't even human'. Consequently, he wants to kill the buffalo to exterminate the Indians. When McKenzie tells of General Sheridan giving medals to buffalo killers, Gilson replies: 'What's wrong with that?'

Merely brutal at the start of the film, Gilson becomes increasingly psychopathic, consumed by his lust to kill buffalo. His pathological hatred of Indians is similar to that of Ethan Edwards (John Wayne) in *The Searchers* (1956), who at one point fires maniacally into a buffalo herd to prevent his Comanche enemies from feeding on them. Yet the ending of *The Last Hunt* has none of the nostalgia of Ethan Edwards' iconic return to the wilderness at the end of John Ford's film, and makes no attempt to rehabilitate Gilson as a sympathetic or noble character. Unredeemed at the end of the narrative, Gilson dies in a snowstorm while waiting for a duel with McKenzie. The winter landscape is a correlative of his own morally cold nature, as he freezes to death with his gun raised, emblematic of his fixed but impotent desire for killing and vengeance. This warweary revisionist western ends without a gunfight.

The implication underlying *The Last Hunt*, that the relationship between the Plains Indians and the buffalo was more harmonious and sustainable than that of the white hunters, has become a commonplace in Hollywood treatments of the issue. Indeed, as the ecological Native American became an ideal for the 1960s 'counter-culture', so Hollywood's representations of Native Americans became more positive. In *Little Big Man* (1970), the Cheyenne are shown stalking the Plains for buffalo in a quiet and dignified way that contrasts sharply to the white hunters who amass buffalo skins for profit rather than for subsistence.[10]

Dances With Wolves (1990) perpetuates the notion that the relationship of the Sioux with the buffalo was ecologically sustainable. When John Dunbar

(Kevin Costner) discovers a field of buffalo dismembered by white hunters, he comments that the people who perpetrated such actions were 'without value and without soul, with no regard for Sioux rights'. The white hunters killed the buffalo 'only for their tongues, and the price of their hides'. This scene is followed by the Sioux hunt in which Dunbar participates, which is represented as a primal ritual, heroic and proud. The buffalo hunt is filmed with an explicitness similar to that of *The Last Hunt*: animals are seen falling with arrows and spears in their bodies. Nevertheless, the context provided for these images signifies the Sioux hunt as different from that of the white hide hunters. The extended chase sequences are accompanied by classic up-tempo, brassy western music on the soundtrack, signifying the hunt as exciting and heroic. Moreover, the sequence climaxes with Dunbar saving a Sioux girl from a charging buffalo by shooting the animal dead. Hailed as a hero by the Sioux, Dunbar then shares the buffalo's heart with a Sioux warrior. In this way, the buffalo hunt becomes part of the movie's discourse on male heroism and redemption, as Dunbar's guilty complicity in the expansion of the American empire is healed through his contact with Native Americans and their close relationship with nature. After feasting with the tribe on the buffalo meat, Dunbar eulogizes the Sioux, his face lit by a sublime sunset: 'I have never known a people so eager to laugh, so devoted to family, so dedicated to each other, and the only word that came to mind was "harmony"'. *Dances With Wolves*, then, shows the Sioux hunting buffalo for sustenance rather than for financial profit, within a ritual context, and in an environmentally harmonious way.[11]

Recent environmental historians have, however, questioned these commonplace notions. Dan Flores, for example, challenges the assumption that the effect of Indian hunting on buffalo populations was minimal, and that the decline in the animal's numbers was due solely to the influx of white hunters after the Civil War. He notes several cultural and environmental factors that contributed to the decline in the buffalo population on the Plains by 1850, including drought, the importation of new bovine diseases such as tuberculosis and brucellosis, and competition with horses over food and water. Crucially for the myth of the ecological Indian, he argues that tribal hunting practices were also a central reason in the animal's decline. In particular, the replacement of subsistence by market hunting, and the Indians' preference for killing cows for their meat rather than bulls, put pressure on buffalo populations. Moreover, the religious outlook of the Plains Indians, contrary to the popular notion still articulated by movies such as *Dances With Wolves*, may have prevented, rather than encouraged, a true understanding of ecological relationships, in that the belief that the buffalo were infinitely abundant and of supernatural origin inhibited recognition of the precariousness of population dynamics, and the role played by over-hunting in their fluctuations.[12] To call the Plains Indians 'conservationists', therefore, requires an extension of the usual sense of the term, as Shepard Krech concludes: 'to brand them conservationists', he writes, 'is to accept that what might have been most important to conserve was not a herd, or an entire buffalo, or even buffalo parts, but one's economically vital, culturally defined, historically contingent, and ritually expressed relationship with the buffalo'.[13]

Such complexities and uncertainties are clearly beyond the interests of *Dances With Wolves*, which, in keeping with movies such as *On Deadly Ground* and *Pocahontas* discussed in Chapter Four, upholds the myth of the ecological Indian for the redemption it offers for the guilt and uncertainties of modernity.

The feather trade and endangered birds: *Wind Across the Everglades*

Screenwriter Budd Schulberg was led to make a movie celebrating the natural beauty of the recently formed Everglades National Park, and dealing with the conservation of its endangered bird species, after a fishing trip there in the late 1940s. *Wind Across the Everglades* (1958) was the result of his uneasy collaboration with director Nicholas Ray.[14]

The story concerns the anti-feather campaigns of the turn of the nineteenth century, when nature enthusiasts, led by local Audubon societies, scientists, sportsmen and ladies' clubs, mobilized around the use by the fashion trade of bird feathers, particularly from herons and egrets, for the decoration of women's hats and dresses. As the film's opening narration states: 'Nearly everyone conformed to the fashion, and few connected its gaudy demands with the plight of the vanishing birds. Only a small army of volunteers, calling itself the Audubon Society, was fighting to save the tropical birds from extinction'.

Bird populations were particularly under threat as hunters shot adult birds for their breeding plumage, leaving their young to starve. Eventually, the federal government, lacking explicit power to protect wildlife, used the indirect means of regulating interstate commerce to halt the trade. The Lacey Act of 1900 banned shipment across state lines of wild animals killed in violation of state laws. Within a decade, the illegal trade had been eradicated.[15]

The film draws loosely for its plot on the murder in 1901 of Audubon Warden Guy Bradley by a commercial bird hunter he had been trying to arrest in the Everglades. Writing of his screenplay in 1975, Schulberg described the young warden as a 'pure knight of the Audubon, [who] gave up his life to save the tropical birds from the savage plume-hunters who were destroying one of the great natural aviaries of the world'.[16] *Wind Across the Everglades* mediates these conservationist concerns by establishing, and then blurring, binary oppositions central to the western genre: wilderness and civilization, official law and outlawry, sacred and profane, human and animal. The narrative begins with high school teacher Walt Murdoch (Christopher Plummer) removing the feather from a woman's hat, with the admonishment: 'How would you like it if this bird wore you as a decoration. . . . When these Indians wear feathers in their hats, we call them savages. And we think we're civilised!'. Murdoch's opening speech already suggests the blurring of oppositions between savagery and civilization, Indian and white, human and animal that will be further developed as the movie unfolds.

Murdoch is subsequently recruited by the Miami authorities as a warden for the Audubon Society to patrol the Everglades and protect its endangered birds from poachers. In this way, he acts to uphold the forces of law and civilization, represented in the movie by the shopkeeper Aaron Nathanson (George Voskovec), who is dedicated to developing the Everglades for agriculture. 'Three million acres of unexplored frontier', he enthuses. 'I see a whole network of canals bringing the wasteland into cultivation'. Yet Murdoch's reply is ambivalent, revealing his desire to preserve rather than develop the wilderness: 'Beyond this river, is an unexplored empire, all right. And now that I've seen it, I want to help keep it an empire. A sanctuary for wild life'. He explains to Aaron that he never got along with 'progress' very well. 'Maybe that's what draws me into the Glades', he continues. 'You see, they're sort of the way the world must have looked on the first day. When it was all water, and then, the first land beginning to rise out of the sea. You feel the life-force in there in its purest, earliest form. Then Cain, and the brothers of Cain, raise their twelve-gauge shotguns and fire into the face of God'. Murdoch is a preservationist, wanting to leave Edenic nature as a separate sphere away from human interference.

Nevertheless, in keeping with the ambivalence described by Richard Slotkin at the start of this chapter, the wilderness hero reluctantly works to establish the very civilisation he distrusts. Despite his desire to preserve the wilderness, Murdoch is aware that his sorties into it provide information for the 'improvement commissioners' in Miami. Moreover, when Aaron tells him he has plans for paving two miles of Flagler Street, Murdoch replies that he is not against the idea, adding that his survey of the poisonous tree he found in the swamp will be useful to the town authorities. As Aaron says to Murdoch's fiancée Naomi (Chana Eden): 'you'll make a civic leader out of him yet'.

However, Murdoch finds himself attracted to his antithesis, the outlaw plume-hunter Cottonmouth (Burl Ives). His band of 'swamp rats' are market hunters who take a sensual pleasure in the slaughter of the birds, and live off nature's bounty, in a life of drink, food, music, sex and violence, the delights of which are summed up by Cottonmouth's catchphrase: 'Ah, the sweet tasting joys of this world!' Cottonmouth is thus a man of nature: 'When I'm three score and ten I'll die here', he tells Murdoch, 'with the seeds of a swamp cabbage in my gut, so a tree'll grow out, and stand on top of me'. As such, he is presented as the literal inverse of Murdoch, first appearing upside down in the lens of the latter's camera. However, as the movie proceeds, the opposition between Murdoch and Cottonmouth is blurred. Murdoch's contact with the wilderness changes him, so that he becomes, as Aaron the storekeeper says, 'almost wild'.

The plot of the movie turns on the classic western choice between duty and desire. Sick with swamp fever, Murdoch refuses to take the advice of his doctor, to settle down in the town and marry his fiancée. Instead, although Naomi wants him to stay, he decides to go back into the wilderness to do his duty in upholding the law by arresting Cottonmouth and bringing him to justice in Miami. However, when Murdoch re-enters the Everglades to arrest the poacher, the tone of the movie begins to change, and he starts to find the primitive, self-reliant masculinity of Cottonmouth's gang preferable to the constraints of civilization. Murdoch proves himself to be 'no pantywaist', as one of the gang approvingly states: 'He holds his liquor, he smokes big black cigars and he's pretty handy with his dukes'. Drunk on moonshine, Murdoch comes to understand why the swamp gang wishes to hold out against the forces of progress. 'No worries, no responsibilities. Wild, like the rest of the wildlife. I'm beginning to see what you've got here', he tells Cottonmouth. One of the gang replies that Cottonmouth 'carries the freedom of the individual to its logical conclusion'. Living in the Everglades is a 'protest' against the 'whole overgrown spider web we call civilisation'. Cottonmouth summarizes their attitude: 'I'm agin' everything, exceptin' the jug'. The bird hunters are anarchic naysayers and self-reliant individualists, upholding a form of primitive masculinity threatened by modernity. Established in the opening narration as blasphemous ('The choicest plumes preserved between the pages of musty books, or sometimes, even a Holy Bible'), the poachers enjoyment of the profane flesh is also represented as seductively attractive.

Ultimately, however, Murdoch rejects Cottonmouth's values because of the damage they are doing to the wilderness he loves. The opposition between Cottonmouth and Murdoch is thus a clash between two different conceptions of nature. Cottonmouth evokes a Darwinistic version of nature to justify the predatory life of his swamp gang. 'We don't need your Ten Commandments in the Glades', he tells Murdoch, 'we do all right with one: eat or be ate'. Cottonmouth uses this argument to justify his slaughter of the birds. 'Sure, I eat the birds', he says, 'the birds eat the fish, and some day, something will eat me, too'. For Cottonmouth, the ecological notion of the interconnected web of life justifies his predatory attitude towards nature.

Murdoch, on the other hand, believes Cottonmouth, named after the poisonous snake he carries in his pocket, to be the serpent in the Eden of the Everglades, his actions a violation of natural order and balance, rather than their embodiment. 'I know a little something about the balance of nature', he tells Cottonmouth. 'I know that fish, birds, snakes, including you, have to eat each other to keep that balance. But wiping out whole rookeries, tens of thousands of birds, with twelve-gauge shotguns, just for some easy money, silly fads, and fat profits. That's got nothing to do with the balance of nature. That's just greed and destruction'. For Murdoch, Cottonmouth's greed and mercenary excess threatens, rather than epitomizes, the natural balance of the Everglades.

The nature footage used in the movie tends to confirm Murdoch's view of nature against that of Cottonmouth. Murdoch's first canoe trip into the Everglades constructs nature as predatory, savage and threatening, but also sublime and beautiful. In a short narrative sequence, a bird eats a fish, and an alligator eats a duck, with a crunching noise on the soundtrack. Then, in a low-angle close-up, two egrets hiss aggressively at the camera, before the film cuts to a shot of Murdoch nervously paddling his canoe. Human beings, in this savage Eden, are thus constructed as intruders in nature. A moonshiner warns Murdoch to get home before dark, because 'they got a little bit of everything in there'. Nevertheless, these scenes construct nature not only as a scene of horror, but also one of sublime beauty. For Murdoch, Darwinian nature is augmented by this more benign perception of nature as a scene of aesthetic pleasure and spiritual contemplation.

In contrast to the nature sequences in which Murdoch is featured, the bird hunting scenes involving Cottonmouth's gang are shockingly excessive and explicit in their violence. Guns are shown actually killing birds, with birds falling dead in the same take. An obviously injured egret is shown walking up a branch towards its nest. These scenes unequivocally signify Cottonmouth as a perverter of the natural order. That he uses a poisonous tree to kill the benign Seminole Indian Billy, who has been driven from his land by the bird hunters, further demonstrates the evil of his actions.

The movie's climactic sequence, when Murdoch and Cottonmouth attempt to row through the swamp to Miami, brings together the two views of nature in an uneasy reconciliation. Murdoch points out the birds to Cottonmouth. 'Can't you see they're beautiful?' he asks the poacher. 'Why can't people just enjoy them, instead of killing them? Just enjoy them and be thankful for them?' Cottonmouth still does not understand: 'I can never get it through my head why you keep taking such chances for birds', he replies. However, when Cottonmouth is fatally bitten by a poisonous snake, he is able for the first time to see the birds as aesthetically beautiful, rather than merely as commercial objects to master and exploit. 'Maybe, maybe you was right', he says to Murdoch. 'Guess I never had a good look at them before'. He then sees a raptor swooping down towards him, and, with an air of defiance, accepts the inevitability of his death: 'Come 'n git me—swamp-born 'n swamp-fattened, tough 'n tasty Glades meat'. Murdoch appears to fight back tears as he watches Cottonmouth die.

The ambivalence of Cottonmouth's death confirms not only his own view of nature as a scene of predation, thereby fulfilling his earlier prediction that 'some day, something will eat me, too', but also, in his final recognition of the beauty of the birds, the more benign view of nature that informs Murdoch's preservationist stance. Cottonmouth's Darwinist version of nature, then, has proved to be only part of the truth: nature is both cruel and beautiful. Both Cottonmouth and Murdoch are, in their different ways, correct.

The ambivalence of this ending confirms Budd Schulberg's own comments on his intentions for his screenplay, which were to create sympathy for the plume-hunting outlaw, as a representative of lost frontier values of self-reliance, even if those values were antithetical to the conservationist beliefs informing the foundation of the National Park. As he put it:

> the embattled, no-use-for lawmen Billy B.'s, enemies of the Park and of the birds, yet impossible to consider as villains: Somehow they belonged to the Everglades as much as the saw grass and the big cypress, the white herons and the black turkey buzzards. . . . Is it our inevitable tragedy that in our praiseworthy efforts to preserve this wilderness, we must tame or drive out the men who were its natural inhabitants? Perhaps, in order to progress, to conserve, to further the constructive ends of society we must sacrifice some original quality of self-reliance. The Cottonmouths must die. And the Walts, the wardens, whose mission is to destroy these Cottonmouths, weep at their passing.

Schulberg also wrote of wanting to achieve a 'complexity of the sort that could make my story of the Everglades not merely cops-and-robbers, warden versus plume hunter, a *Western* moved to the Badlands of the Far South, but a human drama in which right and wrong blend into and overlap each other'.[17] As the above analysis of the film demonstrates, *Wind Across the Everglades* does draws on narrative and character types derived from the western genre, but it does so in a way that, as Schulberg said, complicates the Manichean framing of issues typical of many westerns. Despite this, the movie was advertised in the *Miami Herald* as the 'first story of the people, passions and plunder that swept the 1000 terror miles of the Everglades', and premiered at the Beach Theater in Miami with a parade of Seminole Indians 'in costume'.[18] Unsurprisingly, perhaps, the more sedate pages of *Audubon* magazine failed to mention a film that was not only somewhat ambivalent in its denunciation of commercial bird hunting, but also relished its nostalgic and ribald portrayal of Miami low-life at the turn of the nineteenth century.

Anti-whaling narratives

Herman Melville's novel *Moby-Dick* was first filmed in the silent era as *The Sea Beast* (1926), a movie not for literary purists. Whaling captain Ahab Ceeley (John Barrymore) seeks revenge on Moby-Dick after he loses his leg in his first encounter with the whale, and believes that his fiancée Esther has rejected him because of his disability. In a melodramatic reversal, however, he later discovers that his evil half-brother Derek pushed him into the sea, and deceived him into thinking that Esther had rejected him. So Ahab gives up his quest for the whale, and is reunited with Esther. The feminized values of hearth and home thus triumph over those associated with life at sea. 'Why, Ahab Ceeley, you're crying', declares Esther as her sea captain returns.

The Sea Beast is as uninterested in winning sympathy for the whale itself as it is in producing an authentic version of Melville's novel. Moby-Dick is first described as 'that very white, and famous, and most deadly immortal sea beast who had attacked and slain unnumbered men'. However, despite the subsequent lack of interest in the whale as an adversary to the hunters, the suggestion made by the movie that whaling is linked to madness and emotional frustration in the male hunter is a theme taken up again by later movies, including John Huston's version of *Moby-Dick* (1956), and *Orca* (1977).

In the latter movie, the white hunter Captain Nolan (Richard Harris) is represented as a force outside of nature, who uses crude mechanical devices, such as winches, chains and harpoons, to transform nature from a prelapsarian idyll into a Darwinian scene of brutal struggle. The movie begins with a sequence of orcas in the wild which uses slow motion and rhythmical cutting on their breaching movements to signify the animals as beautiful and graceful, connotations reinforced by Ennio Morricone's music that, in its tone colour and harmonic structure, invokes pastoral conventions. The orcas are part of a pristine wilderness that connotes individual freedom and joyous, sensual play.

Captain Nolan, however, wants to capture a live whale to sell to an aquarium in order to pay off the mortgage on his fishing boat. When he accidentally kills a female orca, who aborts her unborn foetus, the grieving male orca attacks a member of Nolan's crew in revenge. The male orca thus becomes violent towards human beings only after his family is destroyed by the hunter, whose desire to exploit and commodify nature provokes nature's revenge.

The thinly disguised phallic suggestiveness of the hose and harpoon used by Nolan to hunt the whale genders his desire to dominate nature as masculine. Moreover, Nolan's desire is signified, as in Melville's *Moby-Dick*, as an act of blasphemous transgression: hence his exclamations ('Mother of God', 'Merciful God') as he attempts to kill the whale. Like Ahab, Nolan is driven mad and ultimately destroyed by his obsessive quest to master non-human nature.

As Nolan comes increasingly to identify himself with the orca, human protagonist and animal antagonist become mirror-images of each other. Nolan, the viewer learns, lost his pregnant wife and unborn child to a drunk driver. By doubling Nolan and the orca in this way, the movie universalizes the male role of protector of the female, engaging in competitive struggle with other males, as an essential and primitive fact of nature. Nolan sublimates his guilt and bitterness over his wife's death into his sado-masochistic obsession with risk and adventure. The movie emphasizes the uncanny aspect of the growing identification of human and animal with several big close-ups of Nolan reflected in the orca's monstrous eye.

In contrast to Nolan's desire to kill the whale, a decision endorsed by his Native American friend Jacob Umilak, the marine biologist Rachel (Charlotte Rampling) is the voice of ecological concern in the movie, her name recalling the ship in *Moby-Dick* that stands for feminine pity.[19] In the lecture which opens the film, Rachel describes the intelligence of the orca, its status as exemplary parent ('better than most human beings'), and its possession of a 'profound instinct for vengeance'. Marine biology here gives pseudo-scientific justification for the movie's anthropomorphization of the orca, the imputation of the motive of personal revenge to the wild animal recalling similar explanations in the *Jaws* series, which will be discussed in the next chapter. However, in *Orca*, unlike in the *Jaws* movies, the wild animal survives all attempts by the hunter to kill it. In this sense, the movie handles the notion of human mastery over nature more ambivalently than the *Jaws* movies. Indeed, Rachel tries to dissuade Nolan from hunting the whale, arguing for 'its right to be left alone'. Rachel is the voice of sanity, empathy and compassion for nature, a perspective endorsed by the lyric sung over the closing credits: 'My love, we are one'. Like Ishmael in *Moby-Dick*, Rachel also possesses qualities which grant her survival at the end of the narrative, even as the male's desire to master nature ends in his death.

However, the resolution of the movie is ambiguous, in that Rachel is ultimately forced to abandon her non-violent attitude to the orca, when she finally urges Nolan to use his rifle to protect himself from the attacking whale. Nolan's traditional, masculine belief in the necessity of violence as self-defence is thus borne out. The movie therefore combines an uneasy criticism of the masculine desire to master and exploit nature with a partial vindication and ennobling of such desire.

The monstrous whale created by *Orca* was swimming against a cultural tide of anti-whaling sentiment and activism, which had come to public attention a couple of years before the release of the film, when in 1975 Greenpeace first followed Soviet whaling ships off the California coast in an attempt to prevent the hunting of sperm whales.[20] In contrast to *Orca*, later Hollywood movies have unequivocally endorsed this anti-whaling agenda. *Star Trek IV: The Voyage Home* (1986) protested against the Russian hunting of humpback whales, in a science fiction scenario in which the extinction of the whales leads to a threat to the Earth itself from an alien life-form which had been communicating with them. The Star Trek crew return to late twentieth-century America to intervene in history and prevent the extinction of the whales. Captain Kirk (William Shatner) spells out the anthropocentric reason that the movie gives for saving the whales: 'When man was killing these creatures, he was killing his own future'.

In comparison, *Free Willy 3: The Rescue* (1997) advocates the banning of whaling out of respect for the rights of the animals themselves. The orcas in this film are being hunted by illegal American whalers providing meat for markets in Norway, Russia and Japan. Like *Orca*, the movie links whaling to traditional notions of rugged masculine identity. It does so, however, in a way that avoids the equivocations of the earlier movie. The whaling captain John Wesley (Patrick Kilpatrick) takes his ten-year-old son Max (Vincent Berry) on a whaling trip, and tells him of his own father's life as a whaler. Max's grandfather used to call whaling 'God's work', he tells his son, 'because for hundreds of years the whales they killed made oil that filled the lamps that lit the world. Whalers made light. It was work that meant something then'. Now, he adds, whaling means 'you have a right to be what you are. To make a living'. At the start of the film, then, Max's father views whaling as part of his identity as a father and breadwinner for his family.

This identity is, however, shown to be based on violence towards non-human nature. 'Never take your eye off the ball', he warns Max. 'If you do, the sea'll get you, or the whale will'. 'Don't you worry', he adds, 'We'll get him first'. Showing his son how to use a speargun, John recalls Max's fondness for violent video games. Max will be a good shot, he says: 'Kid's been doing target practice his whole life'. However, *Free Willy 3* reveals the father's type of violent, sacrificial masculinity to be outmoded. He is shown to have a paranoid attitude towards nature, and is taught a lesson in environmental awareness by his son, who turns against his father because of the violent nature of his job.

The child's conversion to ecological awareness comes when he falls into the sea, and meets the orca Willy face-to-face, in an epiphanic moment rendered in slow motion, with female vocalese on the soundtrack. Willy does not attack Max, as the boy expects him to do. In place of the harsh, violent natural world assumed by his father, then, Max discovers that nature is feminized and gentle.

As in the earlier *Free Willy* movies discussed in the next chapter, the basis for environmental ethics in the film is the notion that the orca is a 'friend' of the main protagonist, Jesse (Jason James Richter). *Free Willy 3* amplifies this notion into an explicit discussion of animal rights. In an exchange over the morality of whaling, Max's father tells him that it is wrong to hurt people, 'because people have rights'. Whales, on the other hand, 'are animals. They were put here by God for us to hunt. You don't get upset when you eat a cheeseburger do you?' In the end, however, the narrative bears out Max's attitude of respect and concern for the orcas. When Jesse rams the whaling boat, John falls into the sea, and his life is saved by Willy. This action converts the whale hunter, who admits that his son was right after all: whales *are* like people, and therefore have rights too. Max retorts: 'Cheeseburger saved your life'. Orcas, the movie suggests, are deserving of the right to life because of their similarity to, rather than difference from, human beings.

Free Willy 3 ends with the healing of the father-son relationship placed in crisis earlier in the film. 'What am I supposed to do now?', asks the father desperately of his son. 'I'm a whaler. That's what I am'. Max tells him, 'Not to me. You're my dad'. The narrative resolution is amplified by a coda in which Willy's son is born. In this way, the movie ends by naturalizing and universalizing a notion of benevolent fatherhood. 'Willy's going to make a good father, don't you think?', comments Jesse. The doubling of human and animal father is made even more explicit when Jesse names the baby orca Max. John has learned to be as benign a father as Willy, learning from the orca that fatherhood should include gentleness and compassion.

As in *The River Wild*, examined in Chapter Three, the redemption of the father is predicated on the belittling of the female. Female biologist Drew (Annie Corley) tells Jesse and his Indian friend Randolph that she would not have acted like Willy did towards John. 'Some guy tried to kill me,' she says, 'I wouldn't save his life. I would've bit his butt'. Jesse replies: 'Well, maybe he's smarter than we are'. Randolph adds: 'Or more human'. To seal this dressing down of the female, Drew falls victim to the practical joke concerning dirty binoculars that she had inflicted on Jesse at the start of the story. As the father is redeemed, so is the woman hazed. *Free Willy 3*, then, questions the link between masculine identity, hunting and the conquest of nature, without putting male hegemony over women at risk.

The anti-hunting movies discussed in this chapter suggest that the need to preserve wild animals as figures of threatened wilderness has come to challenge the masculine desire to conquer nature through violence. The next three chapters explore the changing symbolic roles played by particular wild animal species in Hollywood movies since the 1960s, and the emergence of conservationist advocacy as an issue in popular American cinema in this period.

SEVEN
North American Ocean Fauna

'This time it's personal': *Jaws* and the great white shark

The 1970s saw a revival in Hollywood adventure movies in which wild animals were viewed as obstacles to human progress. However, unlike the whales and octopuses that were threatening objects needing to be conquered in *The Sea Beast* (1926), *20,000 Leagues Under the Sea* (1954) and *Moby-Dick* (1956), the representation of the great white shark as evil in *Jaws* (1975) coincided with the growth of the modern environmental movement. Indeed, *Jaws* and its many successors may be seen as a backlash against this rising concern for the environment.

From the critical realist perspective outlined in the Introduction to Part Two, all four *Jaws* movies made to date misrepresent the ethology of the great white shark. In the first movie, marine biologist Matt Hooper (Richard Dreyfuss) advocates the theory of 'territoriality', according to which a single 'rogue shark' 'keeps swimming around a place where the feeding is good until the food supply is gone'. The great white shark has 'staked a claim', he says, to the waters off Amity island. This anthropomorphic projection of possessive materialism onto the shark positions the animal as a competitive rival of the human beings who have also staked a claim to the area. The marine environ-ment, according to this theory, is not big enough for both of them.

Yet the marine biology presented in the film is clearly reductive and overly mechanistic. 'What we are dealing with here', announces Hooper, 'is a perfect engine—an eating machine. It's really a miracle of evolution. All this machine does is swim and eat and make little sharks. And that's all'. This Cartesian view of animals as mindless machines positions *Jaws* in opposition to contemporary concerns for the welfare and rights of individual animals.

For experienced shark hunter Quint (Robert Shaw), the great white is equally threatening, though more supernatural. 'Sometimes that shark, he looks right into you', he says, 'right into your eyes. You know a thing about a shark: he's got lifeless eyes, black eyes, like a doll's eyes. He comes at you, he doesn't seem to be living until he bites you'. The dread evoked here is of an uncanny, vampiric creature, ambiguously positioned between living organism and machine, life and death. For both Quint and Hooper, then, eradicating the threat posed by the great white shark can be the only possible option. Both of these constructions of the animal necessitate the conquest and killing of the shark by human beings.

Jaws notwithstanding, contemporary marine biologists have come to question the notion of the shark as a predator of human beings. It is now thought that sharks do not attack humans intentionally, but only when they mistake them for seals or turtles. Sharks, some biologists argue, do not even like the taste of human flesh, preferring other types of food. Moreover, the idea that sharks claim small coastal areas as personal territory has been challenged by new research which suggests that they roam over large areas. If sharks hunt in wide areas, the attempt to cull a rogue shark becomes pointless, as there is no guarantee that the same shark has stayed in the area of the cull.

Although the *Jaws* movies were clearly not intended to be a natural history of the great white shark, marine biologists have argued that they have nevertheless influenced public opinion about the shark in a negative way. Hunting of the great white increased in the late 1970s, to the point where it is now an endangered species. The *Jaws* sequels, of course, show no concern for such issues.[1]

Jaws 2 (1978) was promoted with a slogan that evoked its audience's fears of real sharks: 'Just when you thought it was safe to go back in the water'. However, the sequel makes even less of an effort than the first movie to create a scientifically credible animal. When Brody (Roy Scheider) is afraid that the shark that he killed in the past (that is, in the first movie) has communicated with another, the new marine biologist Dr Elkins reassures him that 'Sharks don't take things personally'. However, the shark's subsequent pursuit of Brody and his family suggests otherwise. In the sequel, then, the authority of the scientist is now in question, and scientific plausibility is rejected completely for mythopoeic resonance. In particular, anthropomorphic subjectivity has been projected onto the shark, in a melodramatic gesture best summarized by the advertising slogan for the fourth movie in the series, *Jaws: The Revenge* (1987): 'This time it's personal'. In this film, when a shark kills her son off the coast of Amity, Ellen recognizes the personal element in the attack. 'It waited all this time and it came for him', she tells her other son Mike. Now a marine biology graduate student, Mike is initially sceptical of such a conclusion. 'Sharks don't commit murder', he reassures her. 'They don't pick out a person'. However, the shark subsequently confirms Ellen's fears, by following the Brody family down to the Bahamas. The authority of the scientist is again undermined. Indeed, Mike's fellow marine biology student Jake (Mario Van Peebles) demonstrates the low level of scientific credibility aimed for by the movie, when he says of great white sharks: 'they spend half their lives looking for food, and the other eating it. They don't care what it is'. The animal is again reduced to a threatening automaton, needing to be controlled and put in its place by human action.

Lee Drummond accounts for the popularity of the *Jaws* series in the psychic release it provided against the difficult strictures of an environmentalist conscience. 'Despairing of ever assuaging the consuming guilt we have been made to feel toward animals', he writes, 'we lust for a little righteous vengeance'.[2] In constructing the shark as a bad object, the *Jaws* movies are a paranoid projection of humanity's fear of, and inability to master, nature, evoking fears of animal otherness as supernatural and uncanny. As Jewett and Lawrence observe, by enlarging the great white shark by over a third the size of any known shark, *Jaws* turns the animal into a 'chaotic and mythic source of malevolence'.[3] As one of the film's previews stated: 'It is as if nature had concentrated all of its forces of evil in a single being'. In all of the *Jaws* movies, this evil, threatening nature is eventually mastered through male heroism, technology and the blood sacrifice of the wild animal. As Ryan and Kellner summarize the movie, *Jaws* 'projects metaphoric fears of the dissolution of community and family as a result of the venality of business and the weakness of traditional authority figures, and depicts the passage to power of a patriarchal savior of the community who dispatches the threat and restores order'.[4]

Commentators on *Jaws* have rightly explored the movie as a punishment narrative obsessed with the dangers of bodily pleasure and of liberated sexuality, especially that of women.[5] In environmentalist terms, the movie may also be interpreted as a criticism of capitalist greed which, while not instigating the threat from nature, certainly compounds it. When in the first *Jaws* film Mayor Vaughn refuses to close down the beaches after the early shark attacks, he cites economic competition as a matter of survival: 'We depend on the summer people here for our very lives. Amity is a summer town. We need summer dollars. If the people can't swim here, they'll be glad to swim at the beaches of Cape Cod, the Hamptons, Long Island'. *Jaws 2* develops the notion that the exploitation of the sea for human use contributes to the catastrophe. The sea front at Amity is being developed for a private condominium complex. Brody's wife Ellen works for the developers, 'selling the good life', as she says, until the contract is threatened by the shark. In both of these movies, then, business profits are threatened by the forces of nature, and the shark attacks can be interpreted as the revenge of nature on its exploiters.

Of the four movies, *Jaws 3D* (1983) comes closest to developing an environmentalist sensibility concerning human relationships with the great white shark, in that it adds the character of benevolent marine biologist Kathryn Morgan (Bess Armstrong), who trains the dolphins and killer whale at Sea World in Florida. Rather than kill a baby shark when it accidentally swims into the marine park, Kathryn wants to preserve it as the first great white held in captivity. However, the greed of entrepreneurial capitalists again provokes the revolt of nature. The manager of Sea World, Calvin Bouchard (Louis Gossett, Jr), acts against the marine biologist's advice, and puts the baby shark on display to the public. When the animal is traumatized and dies, its mother seeks revenge on Sea World.

The baby shark is also exploited by photographer FitzRoyce (Simon MacCorkindale), who wants to film the killing of the mother shark for commercial reasons. 'If we kill this beastie on camera', he tells Bouchard, 'it can guarantee you media coverage'. However, FitzRoyce is killed when he tries to film his encounter with the shark: nature resists being commodified, and takes its revenge on human commercial hubris. At the end of the movie, the control room at Sea World, an enclosed high tech environment of monitor displays and flashing lights, is smashed by the shark attacks.

Despite this quasi-environmentalist criticism of capitalist greed, *Jaws 3D* nevertheless endorses the eventual killing of the mother shark, and is therefore complicit in the very attitudes it criticizes in FitzRoyce: presenting the killing of a shark as a commodified spectacle for paying spectators. After the mother shark is blown up with explosives by Mike Brody (Dennis Quaid), the movie ends with the two Sea World dolphins breaching their pool, happy under the stewardship of their female trainer. These final images serve to take the curse off animal captivity, and demonstrate that Sea World is a benevolent establishment after all, once the human monsters of capitalist profiteering, and the truly wild animal, have been properly expelled.

The fourth movie, *Jaws the Revenge*, also raises initial environmentalist concerns, only to undermine them in an orgy of violence against nature. Mike Brody is introduced as having an environmentalist conscience, having refused to take research funding from the navy because, he says, 'they put bombs on dolphins'. His colleague Jake, in contrast, believes that 'money is money'. This mercenary hubris is again punished, when Jake becomes a victim of the shark. However, Mike's environmentalist concerns are also apparently abandoned when he is forced to join in the hunt to kill the shark. The great white shark, then, remains beyond the pale of ecological conscience. With a fifth *Jaws* movie in pre-production in 1998, the rehabilitation of the reputation of the great white shark remains a distant prospect in Hollywood cinema.[6]

'Friends to man': the dolphin

Gregg Mitman usefully traces the history of the presentation of dolphins as entertainment in American popular culture, key moments of which include the performances in the 1950s of the trained dolphin Flippy at Marine Studios in Florida, opened in 1938, the underwater documentaries of Jacques-Yves Cousteau on television in the same decade, the 1963 release of the children's movie *Flipper* (1963), and the popular television series that followed its success. 'The making of the dolphin into a glamour species within American culture', writes Mitman, 'was, like the making of natural history film, the result of much behind-the-scenes labor in which scientific research and vernacular knowledge, education and entertainment, and authenticity and artifice were edited and integrated into the final scenes that appeared before the public'.[7]

The 1963 film *Flipper* is typical of Hollywood's treatment of wild animals, in its uneasy reconciliation of contemporary science and vernacular knowledge with the presentation of the animal as visual spectacle in the tradition of the circus. At one point, fisherman Porter Ricks (Chuck Connors) teaches his son Sandy (Luke Halpin) scientific facts about the dolphin. A dolphin 'isn't a fish', he informs him. 'They breathe air, and have eyelids, and have babies just like people'. *Flipper* goes on to reconcile this scientific view of the dolphin with three folk legends about the animal which have been in existence since the time of ancient Greece, all of which the narrative repeats: Flipper leads Sandy to a new fishing ground, fights with a shark, and saves him from drowning by carrying him to safety on his back. These old sailors' tales confirm that the dolphin, as the movie puts it, is 'friends to man'.[8] Typical of the Hollywood wild animal movies discussed in Part Two of this book, then, concern for animal welfare focuses on an animal which is seen to be friendly towards human beings and to share attributes with them. In this case, Flipper is acceptable because of 'his' supposed friendliness and intelligence (the actual dolphin used in the film, Mitzie, was female). As the famous title song puts it, Flipper is 'kind and gentle', intelligent ('No-one you see / Is smarter than he'), speedy ('Faster than lightning'), and with a child-like sense of joy ('in a world full of wonder / 'Smiling there under / under the sea').

With *Flipper*, then, the nature-faking pioneered in cinema by Disney's True-Life Adventures continues into the wild animal feature films of the 1960s, with animal subjectivity constructed in such a way that the Other becomes the Same, and animal otherness is denied.[9] At the basis of this anthropomorphization of the animal is a fantasy of inter-species communication that is as scientifically dubious as it is emotionally compelling. *Flipper* constructs this fantasy by making believe that the wild animal is not only able to imitate, but also to understand, the conventional, arbitrary signs of human body-language. The movie thus assumes that behavioural signs in animals that are visually similar to those in human beings signify the same meanings as those human signs: the dolphin's mouth movements are interpreted anthropomorphically as laughing and smiling, the nodding of his head as 'yes' (for example, in reply to the question, 'do you love me?'), and the waving of his fin as 'goodbye'. In other words, the animal appears to do one thing through dramatic context and editing, when, in ethological terms, it is actually doing something with an entirely different meaning.

By disavowing the reality of animal training in this way, *Flipper* was able to construct the relationship between child and animal as a 'friendship' based on mutual empathy and emotional attachment, thereby allowing nature to serve as a basis for children's fable: Sandy's decision to work extra hours to earn fish for his new pet teaches him the Protestant virtues of hard work and obedience to paternal authority. The implications that this fictive treatment of the wild animal has for environmental politics is, however, controversial. Jacques-Yves Cousteau, commenting that the popularity of dolphins lay in what he called their 'built-in smile that has nothing to do with smiling', appealed for the environmental movement to separate itself from such 'exaggerated sentimentalism'.[10] As a scientific ecologist, the priority for Cousteau was on the survival of species and ecosystems, rather than the life of an individual animal. For preservationist writer Bill McKibben, on the other hand, the kind of anthropomorphism evident in *Flipper* remains an effective means of gaining popular support for animal welfare issues. 'Your can of tuna has a little "dolphin-safe" symbol on it', he writes, 'because in 1963 Chuck Connors starred in a film called *Flipper*, which gave birth to a TV show of the same name'. The allusion is to the Marine Mammals Protection Act, passed by Congress in late 1972 in response to humanitarian public concerns over the indiscriminate killing of endangered ocean species such as dolphins, seals and whales.[11]

The ongoing debate over priorities in wildlife conservation, and the relationship of popular representations of wild animals to those debates, is a topic to which we will return. As far as the 1963 *Flipper* is concerned, the model of human–animal relationships it implies is a moderate one in environmentalist terms, based as it is on notions of paternalistic stewardship. In the first part of the movie, nature is constructed in a way that connotes its uncontrollability and hostility to human beings. The Florida Keys fishing community is doubly threatened by the Red Plague, an unexplained ocean parasite, and then by a hurricane that claims the life of fisherman Nick Velakis. The sea, says Porter, 'kills even those who know it'. Put in this context, the beneficence of Sandy's relationship with Flipper mystically heals the relationship between human beings and nature which is fractured at the start of the film. Human and animals thus achieve mutual communication in a narrative of mutual salvation: after Sandy saves Flipper, wounded by a fisherman's harpoon, the dolphin later reciprocates, saving Sandy from a shark attack. Porter, who had earlier threatened to kill Flipper after the dolphin ate his entire catch, eventually learns, as the Production Notes put it, that 'dolphins are very special friends of man and that the riches of the sea should be shared with them'. Paternalistic authority, having been undermined at the start of the film, is thereby restored, as the father learns from his son a new respect for nature.

Flipper, then, is a moderate discourse of animal welfare rather than of animal liberation or rights. Porter initially wanted to set the dolphin free, not because of considerations for its welfare, but because he wanted to save money on feeding him. Freeing the dolphin, he believed, would also teach his son a lesson, as he tells him, 'about losing things, and people you love, like Flipper and Nick Velakis . . . [and] in time, everybody, son'. Nevertheless, the movie ends with Porter granting his son's wish to keep the dolphin as a pet in the family's lagoon. Like Pete, Sandy's tame pelican, Flipper is ultimately granted the subordinate role of the boy's faithful companion animal.

The change that the representation of wild animals has undergone in Hollywood cinema over the last three decades may be charted by comparing the ending of the 1963 *Flipper* with its sequel, *Flipper's New Adventure* (1964), and then with the 1996 re-make. *Flipper's New Adventure* demonstrated a shift towards a more explicitly conservationist agenda, when at the end of the movie Flipper is spared a life of captivity in the Miami Sequarium, even though veterinary biologists from that institution had earlier saved his life. Instead, he is set free within a marine preserve, under the protective eye of Sandy's father, now a newly qualified park ranger. Although this decision was doubtless motivated by the need to generate more story possibilities for the dolphin, the granting of relative freedom to the dolphin gives the second movie a more explicit, but nevertheless still moderate, environmentalist message than the first. The dolphin is to be protected under the paternalistic tutelage of federal wildlife management, rather than kept as a private pet, as in the first movie.

The 1996 *Flipper* is significantly different from its 1963 prototype, in that, at the end of the story, Sandy chooses to set Flipper free to return to the sea. The re-make thus develops a conservationist discourse informed by an animal liberationist agenda, in contrast to the captivity narrative of the first movie, and the welfare narrative of its sequel. The 1996 movie also adds an explicitly environmentalist plot line concerning habitat destruction. Villain Dirk Moran is involved in dumping dioxin from herbicides into the ocean. As Sandy's uncle Porter complains bitterly: 'it costs big bucks to dispose of it properly, but if you can't, just dump it in the ocean'. The familiar narrative of mutual salvation now has an explicitly environmentalist aspect, as marine biologist Cathy has to save Flipper from the effects of toxic poisoning. In return, the dolphin and his relatives later save Sandy from attack by a hammerhead shark, and from the dastardly Dirk.

As already mentioned, the 1996 narrative significantly ends with the release of the dolphin back into the ocean. The Sheriff (Isaac Hayes) tells Porter that keeping a dolphin as a pet is illegal, and informs him that if the animal is not put in a licensed captivity programme, he must be released back into the sea. So Porter is faced with the choice of either taking Flipper to an aquarium or setting him free. The Sheriff delivers what is for an animal release movie the ultimate threat: 'if Flipper is still here on Monday, I will personally see to it that he's jumping through flaming hoops at Sea World on Tuesday'. Despite initial objections to his release from Sandy ('he's used to being fed') and from Cathy ('that animal probably can't fend for itself anymore'), the animal liberationist argument wins out over Sandy's desire to keep the animal as a pet. As Porter tells his nephew: 'he belongs out there'. In the end, Sandy gives up the dolphin, with the words: 'he should go, he belongs with his family'. As in the *Free Willy* movies discussed in the next section, empathy between boy and animal is thus predicated on their shared need for a family, and becomes the central motive for releasing the animal back to the wild. The movie ends with an endorsement of federal environmental law, as the Sheriff announces the conservationist equivalent of the arrival of the cavalry at the end of a traditional western: 'We'll clean up these waters', he announces, 'the EPA's on its way'.

Although this mainstream conservationist agenda is an integral part of the 1996 movie, in another sense the representation of the dolphin has changed little from its 1963 antecedent. For both versions uncritically present the dolphin as visual spectacle, particularly in the sequences in which he performs tricks to a paying audience of Sandy's friends, in representational strategies that continue to be derived from the humanizing tradition of the circus. The 1996 movie's animal rights discourse is also compromised when Flipper is rigged up with a submarine camera to help locate the barrels of toxic waste dumped by Dirk Moran. Cathy had previously worked on the US Navy's experiments with dolphins' powers of echo location, an appropriation of dolphins for military use that the movie presents as unproblematic and benign.[12]

The re-making of *Flipper* as an animal release narrative owes much to the popular formula established by *Free Willy*, the movie that helped to rehabilitate the reputation of the orca for a global cinema audience. It is with the cinematic representation of that animal that the following section is concerned.

'How far would you go for a friend?': the orca

That an environmentalist discourse should be constructed around a dolphin is unsurprising, given that animal's legendary reputation for friendliness towards human beings. More remarkable is the change in the representation of an erstwhile sea monster: the killer whale, or orca. In *Namu, the Killer Whale* (1966), then, producer Ivan Tors extended his repertoire from a dolphin to the idealization of a more potentially fierce animal. Moreover, the movie took up a more explicit position on conservationist issues than the *Flipper* movies Tors had produced a few years earlier.

Nevertheless, the benign sentiments expressed in the movie, in which the orca swims to freedom at the end, were belied by the treatment of the real-life orca who was trained for the film. Discovered snagged in a fisherman's net near the Canadian fishing village of Namu, the orca was captured by Ted Griffin, owner of the Seattle Aquarium, and towed to a cove near Puget Sound, thereby becoming the first orca to be taken alive into captivity. Griffin and Ivan Tors then decided to make a film featuring the whale, in order, as Tors put it later, 'to bring the truth about killer whales to the general public': 'we planned to love him, take care of him, learn from him, and . . . maybe one day we would let him go back to the North Pacific to rejoin his family and relatives'.[13] That loving the animal may have been better expressed by not capturing him in the first place, and allowing him to swim off when his 'family' came looking for him, did not seem to occur to either Tors or Griffin. Nor did the contradiction involved in their capturing of another orca, a female (which they named Shamu and later sold to Sea World in San Diego), to keep the lonely Namu company. Eventually, Namu was transferred to a smaller enclosure in Seattle Harbor, after the lease on the cove expired, and drowned when trying to escape from captivity, a few weeks before the release of the movie. The story of Namu, then, demonstrates that attitudes to animal captivity appropriate to the circus and the zoo still dominated the treatment of the wild animal in the making of *Namu, the Killer Whale*. The fantasy of human freedom projected onto the wild animal in the film was thus still predicated on the captivity of the animal itself.

The movie itself concerns the search by natural scientist Hank Donner (Robert Lansing) for the key to inter-species communication. 'I wish I could understand your language, and you mine', he says to the orca. 'Maybe it'll be something more direct than words. Maybe something as simple as touch'. Hank finally succeeds in establishing a 'language of mutual trust' with which he can communicate with the whale. In an epiphanic moment, he rides on the whale's back, and they swim together, touch and play, to uplifting, sentimental waltz music on the soundtrack. The orca, he concludes, is a gentle, playful animal, which does not deserve its reputation as a 'killer' whale. The neotonous features of the orca's body, its smoothness, rounded contours and simple black and white tonalities, contribute to this sentimental (and scientifically inaccurate) construction of the animal as docile.

The movie guarantees the beneficence of Hank's scientific investigation of the whale by emphasizing its compatibility with Native American wisdom, which also understands that orcas are gentle animals. Hank befriends nine-year-old Lisa, who, since her father drowned in a skin-diving accident, has had recurrent nightmares in which he is eaten by a sea monster. Hank tells her the story of a Native American princess rescued from an earthquake by a killer whale. Though he was the 'biggest, ugliest monster', he tells her, 'he wasn't fierce at all'. The whale eventually changes into a handsome prince who marries the princess. Predictably, the Native American story parallels the events in the wider narrative, as Lisa loses her fear of monsters in the dark, and discovers that the orca, as she puts it, is 'nice'. Namu thus becomes the means for a type of pet therapy that heals traumas within the human nuclear family, while the wild animal is considered acceptable because of its supposed human traits, even though those traits, as in *Flipper*, are again projected onto the animal.

Hank's benevolent attitude to the orca leads him into conflict with local salmon fishermen Clausen and Burt, who want to kill what they view as a threat to their livelihood. Hank admonishes Clausen with an explicit conservationist message against the indiscriminate and ignorant hunting of wildlife: 'I know you, Clausen, I've known you all over the world. You kill the whales, you kill the elephants, you kill the birds. Senseless destruction. You don't know what you kill. You don't even know what you are'. In the brief history of Hollywood conservationist advocacy, after *Wind Across the Everglades* (1958) advocated the protection of birds and *The Roots of Heaven* (1958) advocated the protection of African elephants (as discussed in Chapter Nine), *Namu, the Killer Whale* pleads for the protection of the orca from the savagery of hunting.

At the climax of the narrative, Clausen is still unconvinced of Namu's beneficence, and tries to shoot him. In self-defence, the orca overturns Clausen's boat, and then rescues him from drowning. When the fisherman realizes that the whale has saved his life, his attitude towards the animal is transformed. Compared to the 1963 version of *Flipper*, then, the ending of *Namu, the Killer Whale* endorses a proto-animal liberationist sensibility: the whale swims out into the open sea, where he is joined by a female companion, in a pairing which parallels the future union of Hank and Lisa's widowed mother Kate. 'He might even have a brand new family', Hank tells Lisa, before adding reassuringly: 'If he wants to, if he really wants to, he can come back'. The closing chorus summarizes the movie's conservationist message: 'Live and let live, let nature be your teacher, respect the life of your fellow creature'. This respect for the wild animal is rigidly circumscribed, however, as the movie implies that an orca is worthy of conservation because it is intelligent, docile, and a family animal. As the closing song puts it, everybody knew that the killer whale 'wasn't a killer at all'. *Namu, the Killer Whale*, then, ultimately exaggerates the docility of the animal, in contrast to *Orca*, discussed in Chapter Six, which went too far in the opposite direction, and misrepresented its aggressiveness. In both cases, the otherness of the animal is denied to serve the ideological project of establishing nature as a ground for speculations on human society.

Free Willy (1993) revived the image of the benign orca pioneered by *Namu, the Killer Whale*, while developing a more explicit animal liberationist agenda than the earlier movie. Significantly, too, the movie helped to promote the campaign to free Keiko, the orca used in the filming of the movie, from captivity: an ongoing action that demonstrates a shift in attitudes towards wild animals compared to the treatment of Namu in the mid-1960s. *Free Willy*, then, both draws on, and ultimately undermines, the circus tradition of representing wild animals as spectacle. While the sequences of trained animal performance put the orca on display for the audience's visual pleasure, the plot of the film hinges on Willy's refusal to perform in front of an audience, and the desire of the protagonists to release him into the wild. Such desires were, in retrospect, noticeably absent from the films made by Ivan Tors in the 1960s: significantly at the end of *Namu, the Killer Whale*, the orca is not deliberately released from captivity, but swims out to sea of its own accord.

Nevertheless, despite these developments, the anthropomorphic construction of Willy in the *Free Willy* movies is identical to that of Namu in the 1966 film, and to the various orcas (all called Shamu) in the Sea World theme parks. In all of these cases, orcas perform behaviours which are the product of training, but are presented to the audience as spontaneous and natural. In Sea World, as Mullan and Marvin put it, the orcas 'have been anthropomorphised—not in the crass way of a circus animal whose behaviour is interpreted in human terms, but rather as a creature which shares and exhibits the same deep emotions, feelings, and desires of the human trainers and by implication with the rest of the human observers'. Aspects of human subjectivity, such as free will, intellectual understanding and emotional affectivity (such as friendliness and joy), are thus projected onto the performing orcas. These attributes apparently 'do not come about simply because the whales are being trained, rather they are manifestations of the close personal relationship that the trainers and whales have'.[14] Presented as arising from spontaneous and self-willed activity on the part of the orcas, in reality the orcas are the object of behaviouristic controls, within a circus-like regime of containment and discipline. The orca's body is 'docile', in Foucault's sense of a body 'that is manipulated, shaped, trained, which obeys, responds, becomes skilful and increases its forces'.[15]

The free will of *Free Willy* is similarly connoted by elements of animal behaviour that have been manipulated by the film-makers. As in Sea World, the narrative of the movie disavows this fact by attributing quasi-human subjectivity to the whale. As in *Flipper*, then, the growing relationship between child and animal is signified in terms of mutual understanding, empathy and emotional attachment. These affective relations are summarized in the advertising slogan for the film: 'How far would you go for a friend?'

The anthropomorphic construction of the orca in *Free Willy* also attempts to reconcile science and mysticism. The scientific view of the whale is represented by marine biologist Rae (Lori Petty), who gives Jesse (Jason James Richter) a revisionist lesson in the natural history of the orca. They are 'just hunters', she tells him, living not on human beings, but on fish, and killing their prey solely for food. This scientific view of the wild animal is later reconciled with the more emotional and mystical view of the animal represented by the Haida Randolph (August Schellenberg). Randolph's Haida mysticism embodies the values of a prelapsarian and pre-industrial past, advocating a spiritually contemplative and respectful attitude to the orca, and by extension to nature as a whole. Jesse's own empathetic relationship with the animal is informed by this deep ecological mysticism: at the end of the movie, the successful release of the orca from captivity depends on his ability to leap to freedom over the sea wall in response to the old Haida prayer that Randolph had taught Jesse earlier in the story. The empathy between boy and animal thus repeats the fantasy of total inter-species communication that, as already mentioned, is a recurrent feature of Hollywood wild animal movies. The dualistic separation of humanity from nature is thereby mystically healed. Nature is re-enchanted, and the fallen soul of the white male is saved through healing contact with both the benign animal and Native American wisdom and spirituality. Jesse is, as Randolph tells him earlier, 'one lucky little white boy'. Significantly, spiritual contact with the orca is ultimately denied Rae, the rational scientist, and is available only to the child, with the help of his Native American friend. In gender terms, contact is denied the female, who lacks the deep relationship with the wild animal granted the two males.

The utopian desire in *Free Willy* for a re-enchanted relationship with the natural world may be seen, like similar fantasies in *The Lion King* and *FernGully* discussed in Part One of this book, as an ideological response to the commodification of nature, functioning as a 'green-washing' alibi for the global entertainment industry of which the *Free Willy* movies and their merchandising are a lucrative part. In such a context, it is perhaps unsurprising that the environmental politics of the movie are cautious. Nevertheless, the movie does establish a moderate animal welfare and rights agenda, placing value on the welfare of individual animals. The criteria it assumes for human intervention to protect such animals are their sentience and ability to feel pain. In this way, the movie recalls the liberal-individualist animal rights agenda of philosopher Tom Regan.[16]

Moreover, the film places the blame for Willy's mistreatment on the corruption and greed of the owners of the theme park in which he is kept, who, as Rae explains to Jesse, treat the orca as 'a commodity'. Unregulated capitalism, then, is shown to be the main cause of animal abuse. Despite this, however, the animal rights theme is handled with moderation. Rae angrily points out to the theme park owners that Willy was too big and too old to have been caught in the first place, and should not be kept alone, and in a tank suitable only for dolphins. That she does not object to the captivity of orcas *per se* is further indicated when she explains to Jesse that Willy's fin flops over as a result of his being in captivity: the conclusion she draws is that he needs a bigger tank. Rae thus argues for more investment to expand the theme park site in the interests of animal welfare. In addition, Jesse's decision to free the whale comes about when the marine park owners plan to let him die in order to claim the insurance money. Even though the whale has been shown to be missing his 'family', it is this direct threat to his life that finally prompts Jesse to plan for his release. The threat to the orca's life provides the deadline which gives the narrative its forward impetus, but moderates the movie's animal rights agenda.

Free Willy also has an ambiguous attitude to animal performance. Part of the potential pleasure of the movie for its audience lies in the vicarious sense of mastery involved in watching the orca perform under Jesse's command, in scenes of aesthetic spectacle similar to the presentation of Shamu in Sea World. Nevertheless, the movie does suggest more radical notions of animal liberation and rights: that a wild animal ultimately should not be reduced to a commodified spectacle, kept in captivity, or trained for the purposes of entertainment. The orca, Rae tells the amusement park owners, is not a natural performer, and the narrative turns on the animal's refusal to perform in front of spectators, and climaxes with his release back into the sea. This animal liberationist agenda has been reinforced by the Free Keiko campaign, organized by the Free Willy Foundation established by the Earth Island Institute in 1995, in which concern for the welfare and release from captivity of the individual animal doubtless combines with promotion of the *Free Willy* franchise.[17]

The movie may also be interpreted as favouring direct action on behalf of captive animals. Jesse's heroic action in freeing the whale endorses the natural justice of vigilante action, as the boy's rebellious non-conformism is channelled into a socially worthwhile cause. The reception of the movie was also an occasion for discussions and protests over animal rights. For example, producers Jennie Lew Tugend and Lauren Shuler-Donner led a protest against the capture of three Pacific dolphins for a Chicago aquarium, telling the *Los Angeles Times*, 'It's outrageous in this day and age to even consider capturing these playful and intelligent animals from their families in the wild'. [18] The future of the captive orcas at Sea World was also debated in the context of the movie's release. Although director Simon Wincer chose to distance his movie from criticisms of theme parks such as Sea World, which, he emphasized, 'is certainly nothing like the park in our story', he nevertheless told the *New York Times* that the movie had taught him that orcas 'really belong in the sea'.[19]

Free Willy 2 is less cautious in its animal liberationist message than the first movie, in that its narrative ultimately turns on Jesse's refusal to allow the orcas to be captured 'for their own good', and put on show in an amusement park for profit. The second movie also shows a wider ecological awareness than the first, by placing its conservationist message in the context of the threat posed by the oil industry to the habitat of the orcas and other wildlife in the Pacific Northwest. Jesse's friend Nadine, Randolph's goddaughter, senses that the oil company is violating public spaces. They 'come through here like it's their own private highway', she says. When an oil tanker spills its load into the sea, Nadine's concern extends to the ecosystem as a whole. 'It's not all going to be okay', she cries. 'I mean, birds, otters, seals, this whole cove. It's all ruined'. Jesse also rejects the marine biologist's argument that the whales are being confined in the cove 'for their own good'. He angrily replies: 'None of this is for their own good. Not hurting them in the first place would have been for their own good. Not ruining their home would have been for their own good. It's all such a bunch of bull!' Jesse thereby voices an awareness of habitat destruction as a prime issue in environmental protest.

Judged from the pragmatic perspective of scientific wildlife management, the strategy of individual animal rehabilitation endorsed by both the *Free Willy* movies and the release plan for Keiko may be a sentimental distraction from more important issues such as habitat restoration, and therefore a waste of scarce financial resources. The *Free Willy* movies, however, appeal to a different constituency. Their populist notion of individual animal welfare and rights evokes a deep ecological sensibility that ultimately prefers emotion to reason, spirituality to science, and a desire to reconcile humanity with a re-enchanted nature.

EIGHT
Wolves and Bears

'Everything of yours is good': changing images of wolves

The scene in *Jeremiah Johnson* (1972) in which the mountain man played by Robert Redford is attacked by a pack of wolves prompted Lewis Regenstein to question the actor's commitment to conservation, in that he was passing on 'the false stereotype to millions of viewers'.[1] However, although the stereotype of the savage wolf continues to play a role in Hollywood cinema in this way, from as early as the silent era other popular movies have challenged the association of the wolf with violent aggression, producing instead images of the benevolent wolf reminiscent of the nature writings of William Long.[2] When heroine Faith Diggs (Nell Shipman) is lost in the wilderness in *The Grub Stake: A Tale of the Klondike* (1923), she enters a valley in which 'Dame Nature', as the inter-title puts it, is shown to be not dreadful, as she expects, but Edenic. All of the animals are friendly towards her, including the wolves, which are represented as 'anxious parents' of their playful cubs. In this way, *The Grub Stake* draws on the more positive traditional stereotype of the wolf as a loyal family animal.[3] No longer afraid of wild animals, Faith comes to understand that nature is divine, and that, as she puts it, 'Everything of Yours is Good'. Indeed, when male hero Jeb (Hugh Thompson) sees Faith surrounded by bears, and assumes that his damsel is in distress, his immediate reaction is to go for his gun. But Faith stays his hand: in such a benign arcadia, traditional masculine heroics based on the violent mastery of nature are unnecessary.

In the Depression era, Hollywood made Jack London's story *The Call of the Wild*, first published in 1902, safe for a nervous America by reversing the implications of its source material. The 1935 movie thus mitigates the Social

Darwinism of London's text by playing down its fascination with the wolf as an emblem of aggressive, competitive individualism. Instead, the movie appropriates the wolf for the opposite ideological purpose: to guarantee that nature is based on Christian ethics and bourgeois law and order. The 'call of the wild' in the movie is thus not the triumphal, masculinist initiation into savagery of London's text. Instead, the climactic scene of Buck's entry into the wilderness begins with a long shot of sublime Alaskan mountains, with wolves howling on the soundtrack, before dissolving into a pastoral scene in which the dog is lying peaceably next to a female wolf and a brood of playful cubs. At this point on the soundtrack, the wolf howls are replaced by birdsong. Nature, then, is not wild and lawless, but a domesticated, harmonious social order based on the universal nuclear family. Like *The Grub Stake*, the movie again draws upon the positive image of the wolf as a loyal family animal, and plays down the notion of its savage aggression that so fascinated Jack London. Indeed, that the supposedly wild, wolf-like Buck at the end of *The Call of the Wild* is ultimately as soft and benevolent as the civilized wolf-dog hybrid in the sequel *White Fang* (1936) demonstrates that Hollywood held strong reservations about the implications of a Social Darwinist construction of the natural order in New Deal America. The softened image of the wolf also contrasted with the traditional evocation of its aggression in Disney's *The Three Little Pigs* (1933), in which the animal has been interpreted as a symbol of the Depression, threatening family life until its defeat by the Protestant, Hooverite values of hard work and self-reliance.[4]

The 1960s saw the figure of the good wolf emerge unambiguously in popular American cinema, in keeping with the wider changes in attitudes to wild animals discussed in the Introduction to Part Two. In Disney's *The Legend of Lobo* (1962), adapted from Ernest Thompson Seton's short story 'Lobo: The King of the Currumpaw' (1894), the good wolf becomes a heroic protagonist for the first time in American cinema.[5]

Seton's story drew on his experiences as a wolfer in New Mexico, and centres on a lone wolf, a figure which rose to prominence in the fiction and folklore of North America as wolf populations dwindled at the end of the nineteenth century. The lone wolf usually possessed exceptional size, strength, cunning, ferocity and appetite.[6] It was also the wolf of the Brothers Grimm and Hans Christian Andersen, in whose narratives, as Barry Lopez observes, the 'violence done to the wolf is socially acceptable'.[7] Yet the lone wolf was also a romantic figure in North American popular mythology, as Rick McIntyre explains: 'Americans love stories about courageous loners, fighting for a rebel cause, fated to ultimately die, but dying a good death'.[8] Seton's Lobo was a noble adversary for the western tracker, yet his death was nevertheless considered inevitable: 'It cannot be otherwise', says the narrator, before he captures the renegade wild animal.[9]

Disney's *The Legend of Lobo* sets out to elicit more sympathy for the wolf than Seton's story, in that, whereas the literary text takes the viewpoint of the trapper trying to kill Lobo, spectatorial alignment and allegiance in the movie are with the wolf himself. Moreover, whereas Seton's narrator insists that wolf culling is necessary, the movie makes the animals innocent victims of human aggression and cruelty. *The Legend of Lobo* unequivocally criticizes human beings for waging an irrationally destructive campaign against the wolves. At one point in the narrative, two cowboys shoot dead a cougar they find stalking the wolf cubs. Rex Allen, the folksy narrator, comments: 'There was something about cougars that just naturally brought out hatred in cow punchers. Of course, the big cats were death on horses and other livestock but the resentment seemed more'n that, almost, it was a kind of unreasoning, but it generally took the form of a rifle bullet. Now this went for wolves too'. The lack of allegiance shown by the spectator to the cowboys is reinforced by the refusal of the movie to develop the men as characters in any way. The professional bounty hunter from Texas who attempts to track Lobo down is similarly nameless, while his crumpled black hat and thick moustache are traditional connotations of villainy. In contrast, it is the wolves who are imbued with heroic, humanist virtues, ironically lacking in the human characters. The wolf, the narrator informs the viewer, is an 'intelligent critter'. Moreover, when Lobo keeps his dying father company, he proves that there is 'a kind of nobleness in the wolf'.

The movie places its rehabilitation of the reputation of the wolf within a rhetoric of natural history that challenges the notion that wolves kill cattle out of savagery, rather than simple necessity. The narrator also explains that wolf predation of livestock is a consequence of the encroachment of human beings into their habitat:

> It seemed the hungrier his family got, the scarcer the game got. More and more, civilisation was pushing into these parts, killing off his prey, or crowding it out. . . . When the buffalo were wiped out, it left the cattle as the wolves' only hope of survival. So El Feros took what seemed his rightful share. But the cattlemen weren't sharing with anybody, and from the beginning it was open warfare.

This justification of wolf predation contrasts sharply with Seton's text, in which Lobo's pack kill cattle not merely out of hunger, but wastefully and recklessly, with the fastidiousness of the gourmet and the cruelty of the sadist:

> these freebooters were always sleek and well-conditioned, and were in fact most fastidious about what they ate. Any animal that had died from natural causes, or that was diseased or tainted, they would not touch, and they even rejected anything that had been killed by the stockmen. Their choice and daily food was the tenderer part of a freshly killed

> yearling heifer. ... One night in November, 1893, Blanca and the yellow wolf killed two hundred and fifty sheep, apparently for the fun of it, and did not eat an ounce of their flesh.[10]

As in Disney's True-Life Adventure documentaries popular in the 1950s, *The Legend of Lobo* constructs the wolf as benevolent in order to impress it into service as a moral fable for lessons in Protestant, bourgeois, patriarchal values.[11] Lobo's authority and leadership qualities are given as innate and unquestionable. As the narrator puts it, he 'came from a good line. ... The king of all the hunters / Born to lead the rest'. However, whereas Seton's Lobo holds 'despotic power' over his domain, and possesses a 'ferocious temper', Disney's Lobo is a more benevolent ruler, and more bourgeois than aristocratic. Lobo is, above all, a family animal, as the narrator emphasizes: 'It's a common thing in nature for the male animal to kill his offspring, if he can get to 'em. But the wolf differs. He's about the best parent there is, because he's gentle with his young'. Moreover, the wolf is loyal, and mates for life: 'It's for all time, the male and the female both being devoted to each other and the family'. The Disney movie thus anthropomorphizes the wolf into the acceptable bounds of bourgeois, Christian morality, in a similar way to the nature writings of William Long, and the 1935 version of *Call of the Wild* discussed at the start of this chapter. Teaching the values of self-reliance and resilience, Lobo embodies the masculinist, bourgeois individualistic values associated with the American frontier, values for which the movie nostalgically mourns in the modern, urbanized, bureaucratized America of the early 1960s.

Seton's story ends in pathos and tragedy, when Lobo dies after falling 'recklessly' into a snare, faithfully searching for his dead mate, Blanca. In death, Seton projects human-like emotions onto the animal: 'A lion shorn of his strength, an eagle robbed of his freedom, or a dove bereft of his mate, all die, it is said, of a broken heart; and who will aver that this grim bandit could bear the three-fold brunt, heart-whole?' For Seton, the tragedy of this ending guaranteed its authenticity, as he commented in his 'Note to the Reader': 'The fact that these stories are true is the reason why all are tragic. The life of a wild animal *always has a tragic end*'.[12]

In marked contrast, *The Legend of Lobo* ends with Lobo loyally rescuing his mate from captivity and escaping from the trappers. However, there remains a sense that, although the wolves have won this battle, they are losing the greater war. Their habitat remains under threat from human encroach-ment, and they must move on:

> To your ancestral kingdom man has come to stay,
> So Lobo you must lead your pack and family far away.
> Beyond the distant mountains, you know that there will be
> A place where man won't follow, a place where you'll be free.

The wolves are nostalgic symbols for lost wilderness values of individualism, self-reliance and freedom. But the wolf has earned the right to life by behaving, not like an outlaw, but like an ideal bourgeois American.

The novelty of Disney's portrayal of the wolf as a benign animal may be sensed in the tone of surprise and scepticism in several contemporary reviews. James Powers, writing in the *Hollywood Reporter*, commented that 'it takes a certain amount of courage to cast as a hero an animal whose name is synonymous for cruelty and cunning. The film simply and convincingly explores the wolf character and finds quite a few things to say about the better side of his nature'. Hazel Flynn similarly wrote in the *Citizen-News*: 'Imagine rooting for a wolf. . . . It may perhaps even be applauded in vast spaces such as Siberia and Manchuria, the Arctic or Antarctic where wolves are man's sworn and savage enemies. . . . great is the Disney ability to make even a wolf sympathetic'. However, Bosley Crowther more shrewdly noted the elements of nature-fakery by which the image of the good wolf had been contrived. 'Walt Disney's genii with the cameras, the cutting room shears and, especially, with commentaries and music have managed to outfit a wolf with a fetching suit of sheep's clothing', he wrote.[13]

The Legend of Lobo was not the only movie in the 1960s to recreate the public image of the wolf for an American public unlikely ever to have encountered a real one. *Mara of the Wilderness* (1966) contrasted a mystical view of the wolf with that of scientific wildlife management. Mara (Linda Saunders) is the daughter of a leading biologist who has been raised by timber wolves in the Alaskan wilderness after both her parents were killed by a bear. The hero, Ken Williams (Adam West), is a government scientist, studying for a doctorate in anthropology while working for the Department of the Interior Fish and Wildlife Service.

Kelly, his boss at the Department, voices traditional suspicion of wolves. 'Those white devils are the most dangerous animals on the North American continent', he warns Williams. He then instructs Williams to count the wolves in the park for the purposes of predator control. If the numbers are too high, he should 'thin 'em out'; if too low, he is to do what he can to 'protect and maintain the balance of nature. That's what predator control is for'. However, Williams' first-hand experience with the wolves, under the protection of the wild girl, shows that official wildlife management policy is based on prejudice against the animals. The wolves who have raised Mara are not dangerous, but gentle and sociable. Williams finds himself 'tolerated' by the wolves, as he later tells Kelly; their behaviour advances proof of Mara's late father's scientific theories of 'canine communal instinct' and 'social co-ordinates among vertebrate species'. Typical of the Hollywood environmentalist movie, then, a scientific rhetoric is used to justify nature-fakery.

Significantly, the real 'white devil' in the movie is not the wolf but the illegal trapper, Major William Jarnagan of the Exhibitors Agency, who wants to capture Mara to sell to a travelling show as a freak of nature. An old Fusilier, veteran of India, Africa and Asia, he is the white imperialist hunter, who keeps a Native American as his Man Friday to order about, insult and manhandle, and brutally mistreats the animals he keeps in cages. Jarnagan shows his ignor-ance of and disrespect for wolves when he draws on an inaccurate militaristic analogy to describe them. 'Animals in the wild', he boasts, 'are like troops in combat, and I've handled plenty of both. The bigger and tougher they are, you find the less brains in their skull'.

Yet Mara's gentle, sociable and enchanted nature prevails over Jarnagan's paranoid view. When Williams lies injured in one of Jarnagan's animal traps, Mara is alerted to his plight by the birds, in an act of inter-species communication that saves his life. She then sets Jarnagan's animals free, in an explicit act of animal liberation. *Mara in the Wilderness* in this way develops an environmentalist discourse surrounding the wolf which reconciles mysticism with scientific explanations of nature. Paradoxically, however, this view of nature counters the authority of official scientific wildlife management policies towards the wolf.

A similar combination of mysticism and reformed science may be seen in Disney's *Never Cry Wolf* (1983). The movie is based on Farley Mowat's autobiographical account, published in 1963, of his expedition as a Canadian government biologist sent to the Northwest Territories in 1948 to study the eating habits of the Arctic grey wolf, in order to ascertain whether the wolf was responsible for the decline in the caribou population. The movie adds an anti-hunting message to the book, as government biologist Tyler (Charles Martin Smith) eventually discovers not only that the wolves are not responsible for the decline in caribou, but that they are themselves being killed by hunters for their pelts.[14]

Tyler's contact with the wolves leads him to reject the stereotype perpetuated by the old-timer he meets in the hotel bar on his arrival in the Arctic: 'They'll come after you, son, just for the ugly fun of tearing you apart'. As a revisionist natural history, *Never Cry Wolf* aims to correct such prejudices and myths about wolves. Far from bloodthirsty killers, the animals that Tyler observes are part of a close-knit family structure, with what he describes as 'constant and varied displays of affection' between the male and female. Moreover, instead of the pack battles he had read about, Tyler sees the female assert her position as dominant female with 'challenges and assertions that were mostly symbolic: there was no real fighting'.

Tyler develops the hypothesis that the major part of a wolf's diet is made up of mice, not caribou. In order to prove this, he experiments on himself to see if it is possible to survive eating only mice. This scene is the only exception to the movie's naturalism, which is temporarily abandoned for the sake of anthropomorphic comic relief: as Tyler prepares mice in various ways (on kebab skewers, in soups and stews), there is a comic cutaway to a group of mice, watching in what (according to the context) is taken to be wide-eyed apprehension as they await their turn to be eaten. However, with the exception of this scene, *Never Cry Wolf* largely avoids the sentimentality and anthropomorphism of *The Legend of Lobo* and *Mara of the Wilderness.* Although Tyler gives the wolves human names, the movie nevertheless retains a sense of their otherness. The *mise-en-scène* often places the wolves in deep focus, and, as critic Stephen Hunter commented, the 'two white Alaskan [sic] wolves have a way of suddenly metamorphosing into the bottom or the side of the frame, and they always move with silent, noble grace. Yet they aren't forced into creatures beyond themselves. They are always first wolves, and only secondarily metaphors'.[15] Indeed, Tyler never gets close enough to the wolves for them to become pets or tame companions. This distance from non-human nature leads to respect for its autonomy from the human.

Never Cry Wolf is also a narrative of male self-knowledge, in which wild nature becomes a means for the discovery of an authentic, primitive masculinity repressed in the white American male. In the introductory voice-over, Tyler speaks of his childhood fantasy of finding in the wilderness 'that basic animal I secretly hoped was hidden somewhere in myself'. By the end of the film, he has come to understand his own 'staggering insignificance' in the face of sublime nature, and has rediscovered a sense of wonder, 'like when I was a kid'. This act of self-discovery reaches its epiphany when he runs naked with the wolves as they hunt caribou.

The movie dramatizes different attitudes to the wolf as metonymic for attitudes to the natural environment as a whole. It adds to Mowat's book the character of Rosie (Brian Dennehy), a prospector, entrepreneur and hunter, who owns the mineral rights to a hot spring near the wolves' habitat, where he plans to build a hotel that will attract Japanese tourists. Rosie's presence in nature is destructive, as disturbing as the aeroplane he pilots recklessly into the wilderness. At the end of the film, he shoots one of the wolves Tyler had been observing, and displays its tail on his aeroplane as a trophy. The addition of Rosie to Mowat's text, the author himself commented, allowed director Carroll Ballard to concentrate 'all the adverse qualities of human beings into one person', in order for the film to make its central point, 'which is that humans are the really bloody species on this planet'.[16]

In contrast to the destructive attitude of the white hunter and entrepreneur, the movie constructs the traditional Inuit view of nature as ecologically benign. In doing so, it sets up an opposition between traditional and modern practices among Native American peoples. Whereas the old shaman Ootek (Zachary Ittimangnaq) successfully maintains traditional ways, his adopted son Mike (Samson Jorah) compromises with, and is dependent upon, the modern capitalist exploitation of nature. For Mike, 'wolves mean money', a single wolf pelt bringing him $350 to feed his family or buy a snowmobile. Like Rosie, Mike also shoots the wolves that Tyler has been observing, and is last seen smiling contentedly, his formerly toothless mouth transformed by a new set of dentures. 'This thing that's happening is too big for you', he warns Tyler. 'It's a question of how you survive it. The survival of the fittest'. The figure of Mike signifies the compromises made by the Inuit to survive in a competitive modern environment.

Ootek, on the other hand, remains untainted by capitalist values. His exceptional knowledge of, and respect for, the wolves is derived from his symbiotic, non-destructive way of life. Significantly, it is the Inuit creation myth recounted to Tyler by Ootek's wife that provides him with the key to his discovery of the true scientific ecology of the wolf. According to her story, after the Inuit people had over-hunted the caribou for food and clothing, leaving only the sick animals behind, the wolves were born to kill the sick caribou, thereby strengthening the herd and enabling the Inuit people to eat. This folk intuition about the balance of nature inspires Tyler's scientific hypothesis about predator–prey relationships. As in *Mara of the Wilderness*, then, a mystical view of nature is reconciled with a scientific one, in a way that guarantees science as benevolent.

The assumption made by *Never Cry Wolf* that predator–prey relationships are naturally harmonious, although a commonplace of radical environmentalism, has been challenged by recent scientific ecology. Daniel B. Botkin concludes that there is little evidence of constancy and balance in such relationships.[17] Moreover, the assumption that the religious view of nature held by the traditional Inuit people is itself ecologically benign has also been challenged by recent anthropologists. Eskimos, note Robert Ornstein and Paul Ehrlich, 'generally reject the tenets of conservation, again because they believe that animal spirits, not population dynamics, control the future supply of prey'.[18] From this perspective, *Never Cry Wolf* continues to idealize the ecological role of both the wolf and the Inuit.

Disney's version of Jack London's *White Fang* (1991) employs the figure of the good wolf for a typical morality tale, in which the adolescent tenderfoot Jack Conroy (Ethan Hawke), a character added to London's text, enacts a Horatio Alger story of gold prospecting, at the end of which, through hard work and perseverance, he finds wealth in his late father's mine. As in the 1936 version, the Social Darwinism of London's wolf is sanitized in a way that recalls the humanitarian nature writings of William Long. Indeed, producer Marykay Powell's comments on the movie's representation of the wolf are highly reminiscent of Long: 'What's so very important about this movie is how it shows that wolves are really joyous creatures of nature. Though they have to stalk prey to survive, they are not creatures to be feared'.[19]

Nevertheless, the movie also attempted to be faithful to some of the scenes in Jack London's text in a way that proved controversial with animal rights groups, who saw one episode in particular as problematic: namely, the scene in which two coffin bearers (London's characters Bill and Henry are renamed Alex and Skunker in the movie) carry their dead companion Dutch into the mountains for burial, and are attacked by hungry wolves.

The movie adds two elements to London's text which show both Alex (Klaus Maria Brandauer) and Skunker (Seymour Cassel) exhibiting a generous and trusting attitude towards wolves. Alex tries to reassure his companion by explaining that the wolves are only interested in their dogs, not in the men themselves. Wolves, he says, 'won't jump a man, unless hunger's got them totally crazed'. Skunker then tells of his childhood spent with his wolfer uncle in Montana, who would make him shoot wolf puppies for bounty hunters. 'I hated doing that', he tells Jack. These additions to London's text demonstrate the movie's awareness of the history of human–wolf interactions in America, and its desire to revise apparent misconceptions concerning the animals.

The rest of the sequence is, however, closely derived from London, and for that reason ran foul of animal rights advocates. The wolves send White Fang's mother to the campfire to act as a decoy, so that the rest of the pack can attack the men's dogs. In the movie, Skunker is shown shooting at the wolves, and then, off-screen, is killed by the pack. When the black wolf later leads an attack on Jack and Alex, they are rescued at the last moment by two fellow prospectors.

In its dramatic context, this sequence confirms Alex's earlier judgement that wolves only attack human beings when they are 'totally crazed' with hunger. Nevertheless, the scene was criticized by the Humane Association as an 'antiwolf statement'.[20] Activist pressure subsequently prompted Disney to cut out a shot in which a wolf was shown directly attacking Skunker. His death, as noted above, was instead implied in off-screen space. To further guarantee the benevolence of the wolf, the studio also added a disclaimer to the start of the film, written by the animal rights organization Defenders of Wildlife:

> London's *White Fang* is a work of fiction. There has never been a documented case of a healthy wolf or pack of wolves attacking a human in North America. Because wolves were systematically eliminated throughout most of the United States during our early history and continue to be persecuted today, a nationwide effort is underway to reintroduce wolves into wilderness areas and insure their survival for generations to come.

The assertion that no human being has been attacked by a healthy wolf in

North America is a commonplace in radical environmentalism, but is contested in Barry Lopez's seminal study *Of Wolves and Men*, in which he argues that healthy wolves *have* been known to attack Indians and Eskimos.[21] Whatever the truth about wolf behaviour, the figure of the good wolf remains a contested one in the American West. The Mountain States Legal Foundation, a Denver-based pro-ranching organization, asked Disney to remove its disclaimer from the film, arguing that 'while uncommon, attacks by wolves upon humans have, indeed, occurred, both in the early years of the settlement of this country and more recently'.[22] The organization cited two recent cases of wolves attacking children, one fatally.

The controversial scene of the wolf attack in *White Fang* is, in any case, uncharacteristic of the rest of the movie, which tends to suppress signs of lupine aggression in order to construct the animal as benign, in a typically sentimental Disney narrative about the friendship between a white boy and his tame animal companion. The 'wildness' of the wolf, as in *The Legend of Lobo*, is sufficient to connote a nostalgia for a masculinized frontier of individual freedom and potentiality, but is strictly placed within the bounds of bourgeois decency and decorum.

The film's sanitizing of the wolf can be seen in the opening sequences, which depict a pack of wolves chasing a fluffy white rabbit. The excitement and ferocity of the chase is suggested by rapid pans of the wolves' faces in telephoto close-up. However, in a move presumably designed to satisfy observers from the American Humane Association, the sequence is edited so that the wolves and the rabbit are never seen together in the same shot. Nor is a kill explicitly shown. The effect of this treatment is to mitigate the violence of nature, and to sanitize the predatory nature of the wolf. The hunting sequence then cuts to the wolf mother feeding her cubs, accompanied on the soundtrack by soft, lyrical flute music, in a scene that signifies familial love and tenderness. In this abrupt transition from wilderness to pastoral, the goodness of the wolf is reinforced.

As in the 1936 version, then, the movie again mitigates the Darwinistic implications of London's story, in which nature is a scene of cruelty, aggression and mercilessness. Disney's animal acts from the start like a tame dog, far removed from the ferocious killer in London's text. Nor is London's wolf family the cosy nuclear family of Disney. The movie omits the violent competitive courtship of White Fang's mother, described by London as the 'love-making of the Wild, the sex-tragedy of the natural world that was tragedy only to those that died'. In London's story, White Fang's mother Kiche is sold to the Native American Three Eagles. When she returns home, she no longer remembers her son, and attacks him. In the movie, White Fang's mother is wounded in the confrontation with the coffin-bearers, and returns to the den to die. Predictably, the Disney version prefers pathos and the projection onto nature of human nuclear family values to the violent, sexual wolf of London's story. White Fang is thus introduced to the viewer as an orphaned cub howling plaintively and appearing to shake with grief. He then becomes a comically vulnerable figure in a pastoral landscape: he falls into a hole in the ice, tries unsuccessfully to catch a fish, and eventually is caught in the trap belonging to the Native American Grey Beaver (Plus Savage). Again, the savagery inherent in London's wolf-dog is omitted. In the literary text, the cub kills a ptarmigan, 'thrilling and exulting in ways new to him and greater to him than any he had known before'.[23]

However, the movie is closer to London's text in its animal welfare message. In both texts, Beauty Smith is a brutal, maniacal abuser of animals. In the film, when Smith (James Remar) is admonished by Jack for kicking the cage in which he is transporting his St Bernard dog, he asserts a right of ownership and permission over the animal: 'Why not? He's mine!' Jack later discovers a more convivial, mutual relationship with animals than the abusive desire for mastery exhibited by Beauty Smith.

The major difference between the book and the film is that in the latter White Fang is, as the Production Information makes clear, 'turned into a savage, angry beast' only when beaten by Beauty Smith.[24] In London's story, as already noted, White Fang begins as aggressive and savage, and remains so until he is successfully tamed by his 'Love Master', Weedon Scott. In the film, Jack rescues White Fang from Beauty Smith, nurses him back to health, and tames him through patience, kindness and perseverance. Eventually, in an extended epiphanic sequence accompanied by climactic string music, reminiscent of a similar scene in *Dances With Wolves* (1990), the wolf eats out of the boy's hand, in an image of inter-species communication that mystically heals the division between human beings and non-human nature.

Nevertheless, the relationship between human being and animal ultimately subordinates the latter. White Fang becomes Jack's working dog, pulling his trolley in the gold mine, and a faithful pet, who rescues his master in Lassie-fashion when he is trapped in the mine by an explosion. (This scene reworks London's story, in which White Fang's complete domestication is proved when he is wounded protecting the property and person of his master, Judge Scott, from an escaped criminal bent on revenge.) Barry Lopez comments that 'insofar as the wolf became a dog, a pet, or a draft animal in someone's sledge harness he, too, was accepted'.[25] The wolf in *White Fang* is thus valued for his intelligence and usefulness to humans, rather than for being fully wild.

The ending of the movie underlines its ambivalence towards the wildness of the wolf. When Jack becomes co-partner in a hotel that Alex intends to open in San Francisco, the latter advises him to release the wolf back into the wild. 'He will be miserable in the city', he tells him. 'He has to run free.

That's his magic'. However, when Jack decides to stay on in Alaska and work his gold mine, White Fang returns to him, their relationship renewed. Ultimately, then, Disney's *White Fang* wants the wolf both ways: as a domesticated companion, and as a signifier of the romantic individual freedom associated with the wilderness.[26]

The reintroduction of the timber wolf to Yellowstone Park in 1994 was a triumph both for a popularized form of scientific ecology and for modern, largely urban constructions of the wolf as an integral part of American national identity.[27] The Hollywood movies discussed in this chapter, from *The Grub Stake* to *White Fang*, have played a vital part in the rehabilitation of the reputation of one of the most persecuted wild animals on the North American continent. A similar cultural process may be seen in Hollywood's treatment of another charismatic American predator, the bear.

'Born tame . . . made wild by people': changing images of bears

In Disney's *King of the Grizzlies* (1970), based on Ernest Thompson Seton's story 'The Biography of a Grizzly', a rancher, Colonel Pierson (Chris Wiggins), is building an 'empire' based on cattle, logging and mining, which encroaches into the habitat of a lone grizzly bear. The movie is concerned with reconciling what it sees as the necessities of economic growth and development with a preservationist concern for the bear.

The movie represents Wahb the bear as a territorial rival of the rancher: he is a 'real boss bear', for whom 'by right of might, anything and everything was his for the taking'. The bear creates problems for the rancher, pushing over trees and fences, and eventually causing a cattle stampede. As a 'grizzly king', the narrator explains, Wahb is 'accustomed to going where he wanted to go, whether the way was barred or not'. Nevertheless, the narration is careful to limit the aggression shown by the bear, in order to maintain audience sympathy for the animal. 'Now, there's no denying that all grizzlies are unpredictable and potentially dangerous', the narrator announces. 'Wahb was no exception. But today he was in a particularly good mood, a lot more cunning than cantankerous'. Audience allegiance to the bear is further elicited when the rancher is described as 'the first invader' to appear 'in his domain', and begins to track the animal down in order to kill him.

The Cree Indian Moki (John Yesno) has divided loyalties, being both ranch foreman and 'spiritual brother' to the bear. Moki believes that the Great Spirit gave the mountains to the bear for his hunting ground, and therefore wants to resolve the conflict over territory peacefully. Following the Colonel in his hunt for the bear, Moki stops the bear from attacking them by talking to it, and then prevents the rancher from shooting the animal. The bear makes a mark on a tree on the boundaries of Pearson's land to show the limits of his territorial claim. The Native American man has thus become the go-between between the rancher and the bear, reconciling their opposing territorial interests. Although the bear has, in effect, backed down, and deferred to the rancher's interests, the closing narration disavows this power relationship between white man and wild animal, by stating that Wahb would now 'take his rightful place as King of the Grizzlies, and he would reign supreme in this high mountain wilderness for all the years of his life'. The movie's preser-vationist ideology thus attempts to reconcile the competing interests of human beings and wild animals by confining the latter to a separate sphere. In this way, the grizzly bear can be preserved as a symbol of wilderness for a growing capitalist society, while not affecting the expansion of that society.

The emphasis in Disney's *King of the Grizzlies* on compromise and reconciliation between competing interests, ultimately under the hegemony of the white American male, was also a feature of the same studio's *The Bears and I* (1974). 'Somewhere I'd heard that wildlife is born tame, and made wild by people', comments Bob Leslie (Patrick Wayne), revealing again an attitude to non-human nature reminiscent of nature writers such as William Long. The movie contrasts Bob's attitude to wild animals with that of Sam Eagle Speaker, the drunken Indian rabble-rouser who is responsible for orphaning a family of bear cubs when he leads a group of sportsmen to kill their mother. The essential tameness of the bear cubs adopted by Bob Leslie makes them suitable objects for his stewardship. 'I don't really want them too tame', he tells the official from the Department of Parks and Recreation. 'Just enough to be friendly'.

The bears are accordingly represented in a similar way to Disney's True-Life Adventure movies, with orchestral music closely synchronized to their movements: up-tempo brass when they are fighting, a plodding bassoon motif when they are walking.[28] Dubbed over the music are plaintive cries, comical and child-like, signifying the bear cubs as mischievous children. When Bob finally wins over the trust of the cubs, he comments that 'in some mystical sort of way, a bond was formed. . . That night, for the first time in my life, I had a family'. Nature is once again re-enchanted in the Edenic North American wilderness.

Like its pioneering antecedent *Born Free* (1966), which will be discussed in the next chapter, *The Bears and I* is a release narrative, in which Bob retrains the animals to survive in the wild. He explains his protective role as follows: 'The next day, I began my new career as a mother bear. I planned to do only what she would have done: teach the cubs to be self-sufficient'. Yet Bob's nurturing relationship with the bear cubs leads him into conflict with the Native American tribe, who believe that it is better for a bear to be killed than to be captured alive. In a narrative that does not appear in Robert Franklin Leslie's memoir, on which the film is based, Chief Peter A-Tas-Ka-Nay (Chief

Dan George), father of Bob's friend killed in the Vietnam war, maintains that Bob should set the bears free immediately, telling him that he has angered the Great Spirit by keeping the bears, and predicting bad things for his people as a result. Bob comments: 'I was sorry that old Peter was angry. But Indian superstition wouldn't change my plans for the cubs. Not 'til they were ready, anyway'. The movie thus rejects traditional Indian attitudes to nature, endorsing instead the paternalistic authority of the white male, now sensitive enough (and feminized in his role, as he puts it, of 'mother bear') to want to conserve, rather than simply dominate, non-human nature.

In celebrating Bob's paternalistic stewardship of nature in this way, the movie departs significantly from Leslie's memoir, which is far more ambivalent regarding his adoption of the bear cubs. In the book, over-habituation proves to be fatal for two of the bears, who are killed by hunters because they are too trusting of human beings. 'I was still wondering', writes Leslie, '—as I had indeed wondered throughout the winter—how much better off the bears would have been had I not interfered'. When he finally comes to release Scratch, the last bear under his protection, Bob has to beat it with a willow switch in order to leave 'the thought of betrayal' in its mind, and thereby destroy its trust in human beings.[29]

The movie takes up the reluctance of the final bear to leave his human protector, and hence the necessity of Bob using a rope to beat him away, but it does not repeat the misgivings that Leslie has in the book over whether he should be protecting the animals at all. In the movie, after two of the bears are released successfully to the wild, Patch, Bob's favourite cub, keeps returning, until he is wounded by rebel Indian Sam Eagle Speaker. The Chief refuses to heal the bear's wounds. 'If a bear dies', he tells Bob, 'it is the will the Great Spirit'. However, the white man's viewpoint is again given greater authority than that of the Native American, and is presented as unproblematically humanitarian in its caring attitude to the individual animal. 'The Great Spirit had nothing to do with this', Bob tells the Chief. 'It was Sam Eagle Speaker and his rifle'. He then invokes the memory of the Chief's dead son, to teach the father a lesson in respect for animals: 'You're going to let that bear die for no reason. Larch told me his father had much wisdom. But right now I don't buy that'. Accused by the Chief of lacking respect, Bob adds that there is 'one thing I do respect: life. It's all that matters. Your son and I learned that on the battlefield. And let me tell you this. If Larch were here right now in your place, he'd help his brother the bear'. So the Chief changes his mind, and decides to help Bob heal the bear, making Bob promise in return that if the bear comes to him again, he will drive him away.

At the end of the movie, when Patch returns, Bob drives him away with a rope. 'That was the hardest thing I'd ever had to do', he comments. 'But once it was done, I knew it was right. The bears belong to the wilderness now, and they'd be safe in the new National Park'. He then announces his decision to join the Forestry Service. The movie thus ends by reconciling official wildlife management practices with Native American wisdom, with both seen to guarantee the bears their freedom. Bob's final assurance that the bears will be 'safe in the new National Park', does, however, ring hollow, in the light of Lewis Regenstein's comments on the damage done to the bear population in Yellowstone National Park by official management policies in the early 1970s, when *The Bears and I* was being made. In particular, the sudden closure of the garbage dumps on which the animals had become dependent for food, without the provision of an alternative food supply, together with the enforced 'removal' and relocation of any bears who subsequently entered campgrounds or developed areas, contributed to a further decline in the species, rather than to its rehabilitation. In the light of this contemporaneous history of official mismanagement, the reassuring promise as to the bears' future security made at the end of *The Bears and I* seems complacent and misguided in the extreme.[30]

In keeping with the Manichean view of nature that has been a recurrent feature of the popular fictions explored in this book, the benevolent grizzly bear of popular cinema also has its alter-image, in films which demonize the animal as a savage monster fit only to be destroyed. The post-*Jaws* cycle of wild animal monster movies included the grizzly bear in its demonology, in both *Grizzly* (1976) and *Claws* (1977). In the latter film, a rogue grizzly bear is eventually hunted down and killed by a logger and former hunter Jason Monroe (Jason Evers). As in the *Jaws* sequels, the motive of personal revenge is projected onto the animal to explain its attacks on human beings, even though, from a realist perspective, as Lewis Regenstein shows, fear and self-defence are more plausible explanations for bear attacks on human beings.[31] The wild animal thus again becomes a scapegoat onto which human self-hatred and fear is projected.

The aggression of the grizzly bear in *Claws* is provoked by an illegal guide who neglects to track down and kill the animal when it was wounded by hunters. However, an alternative explanation, based on Native American legend, puts the blame for the bear's aggression on Jason's hunting past. When Henry, his loyal Indian friend, saves Jason from the bear, and is later killed himself, the Commissioner tells of the Indian legend of an 'evil spirit that can take on any form, bear, wolf, an eagle, even a person. It's a kind of punisher. . . . Henry kept the devil bear from doing away with Jason there. And Henry's people didn't like what Henry had done. You see, Jason has killed a lot of bears, and they think it's only right that a bear should kill him'.

Whatever the explanation for the bear's aggression, whether rational or supernatural, the movie does not elicit any sympathy for the animal itself. Instead, as in the *Jaws* movies, the wild animal is signified as inherently monstrous. Moreover, the central dramatic function of the animal is to provide the object against which patriarchal masculinity can be tested and rehabilitated, particularly against the threat posed by female independence. Deserted by his wife Chris after being maimed by the bear, Jason is eventually followed into the mountains by his wife, seeking his forgiveness. After Jason has killed the bear, he and Chris are reunited. In *Claws*, then, wild nature is again signified in a traditionally patriarchal way as a proving ground for the reassertion of white male hegemony. The grizzly bear plays the role of disorderly and savage antagonist to be overcome through justified and regenerative violence in the restoration of civilized order.

The exaggerated aggression projected onto the grizzly bear in *Claws* went against the trend for more sentimental constructions of the animal established by Disney's *King of the Grizzlies* and *The Bears and I*. However, the figure of the benevolent grizzly bear reappeared with the French film *The Bear* (1988), based on James Oliver Curwood's story 'The Grizzly King', and endorsed by the World Wildlife Fund. More recently, the process of idealization and humanization has been extended to the polar bear, in the children's outdoor adventure movie *Alaska* (1996). Here the polar bear cub, like the orcas Namu and Willy, is represented as playful and harmless, in another story in which contact with a benign wild animal is therapeutic for a broken family relationship, and the predatory nature of the real animal is denied. Indeed, the only threat to the Edenic wilderness of Alaska comes from the illegal hunter Perry (Charlton Heston), who is trying to capture a live polar bear cub for his clients in Hong Kong. Perry has a Darwinian view of nature. 'Magnificent creature, the polar bear', he tells his assistant. 'Nature's perfect carnivore. Adapted to the most hostile climate on earth. . . . He is also, along with the leopard, one of the few animals that hunts man. How does it feel, Mr. Koontz, to realise that you are no longer on top of the food chain?'

According to the white hunter, young people 'know nothing of the real world. They can't conceive of the true brutality of nature'. However, the movie confirms the more benign view of nature of its child protagonists, Sean (Vincent Kartheiser) and Jessie (Thora Birch), and proves Perry wrong when the polar bear cub whom the children set free proves himself to be friendly (that is, tame) and possessor of a Lassie-like loyalty to his human companions. The benevolence of nature is again guaranteed by the presence in the film of ecological Native Americans, for whom the bear is a spirit guide. Ben (Gordon Tootoosis) tells Sean to 'trust the bear'. The subsequent narrative confirms this mystical view of the bear, as the animal leads the children to their missing father, injured in a plane crash in the mountains.

Alaska ends with the release of the bear cub back to the wild, typical of the recent cycle of conservationist movies already discussed in this part of the book. The good wild animal has again served its function of restoring the failing white middle-class family, as Sean is transformed from an alienated and resentful adolescent to a mature, responsible adult. 'Thanks for helping me find my way', Sean says to the bear as he gives him his freedom. Again typically, this restoration of masculine identity is predicated on the subordination of the female, as Jessie, Sean's sister, who begins the journey as a resilient leader, is gradually displaced by her brother. Eventually, the son saves his father on the mountainside, when Jessie is not strong enough to hold the rope. Implausibly, the polar bear cub adds his own weight to the rope to save them. Benign nature saves the American family, in the guise of a docile bear, which, like the benevolent dolphins, orcas and wolves discussed in the last two chapters, signifies both a popular interest in the conservation of endan-gered species, encouraged by Hollywood environmentalist advocacy, and a symbolic preoccupation with questions of American social, cultural and national identity.

NINE
African Wildlife from Safari to Conservation

Safari narratives and the conquest of nature

Hollywood outdoor adventure movies have largely promoted a colonialist vision of Africa as a new frontier for European and American interests that was central to the formation of white American national and imperial identity. Until the early 1960s, wild animals tended to be represented in these movies as a savage and primitive Other, and were either trophies to be won, as in *The Snows of Kilimanjaro* (1952), or obstacles to be overcome in the pursuit of wealth or power, as in *King Solomon's Mines* (1950). In both cases, the encounter with wild nature often culminated in the violent death of the animal.[1]

Hollywood drew on popular writers such as Rider Haggard, Edgar Wallace and Edgar Rice Burroughs for this colonialist view of African wildlife. Cinema was an extension of the zoo, as a contemporary reviewer of *Trader Horn* (1931) made explicit: 'It is like a rapid trip through some comprehensive zoo where, for added thrills, the animals are occasionally permitted to get loose and bite each other'.[2] Such movies eulogized the mastery of the white hunter over African nature.

Although the killing of wild animals was central to safari movies, they nevertheless tended to value wild animals higher than the native peoples also encountered by the white hunters. The big game animal was a worthy, noble adversary, sublime and dangerously beautiful, whereas indigenous peoples were merely threatening savages or menial assistants.[3] The colonialist film denied African people agency and voice, and homogenized, simplified and ridiculed them, reducing them to objects of imperial control and mastery. Wild animals, on the other hand, were identified as innocent of the sinfulness associated with dark-skinned people, and victims of forces of progress deemed inevitable. The killing of an animal could, therefore, elicit pathos and regret; African people, on the other hand, elicited no such emotions in these films.

The *Tarzan* series, which began with the silent *Tarzan of the Apes* (1918), demonstrates these tendencies. The first talking Tarzan movie, *Tarzan, the Ape Man* (1932), involved a journey by white ivory traders into an elephants' graveyard considered sacred by the local African tribespeople. Jane (Maureen O'Sullivan) is attracted to the sublimity of the animals they encounter on the safari. The hyenas, she tells Harry Holt, make a 'horrible noise', yet are 'part of it all'. The roaring of the lion is 'grand, so proud, so fierce, yet so infinitely lonely'.

When the party encounters hippopotami in a river, Harry confesses: 'I'm afraid we're intruding', and remarks that he is 'willing to detour'. 'After all, they live here', he adds. Jane notices a mother hippo with her baby, and believes that the animals are 'just curious'. Nevertheless, despite these sentimental concerns for the animals, the killing of the hippos is deemed necessary for the continuance of the safari. The sense of regret for intruding on pristine nature is tempered by a more fundamental acceptance that the destruction of wild animals is an inevitable part of the adventurers' progress.

The figure of Tarzan is a focus for this ambivalent attitude to African animals. Although the jungle is represented as a scene of Darwinian predation, the ape-man is also able to elicit communication and cooperation between species. For example, when he is wounded by Harry Holt, an elephant carries him home and brings him water, while the monkeys fetch Jane to nurse him back to health. However, despite being a man of nature who empathizes with the wilderness in this way, Tarzan nevertheless works for the very forces which aim to conquer it. Ultimately, his empathy with wild animals serves the interests of the white colonialists, in helping him achieve his main purpose: to save the white explorers from Africa's hostile animals and people.

Tarzan, the Ape Man contains several sequences in which wild animals are positioned for the spectator's allegiance in a way denied to the African people in the movie. For example, when the whites are captured by pigmy bushmen, Cheetah the chimpanzee fetches Tarzan, and the spectator is aligned with the animal as he is chased by a leopard and a lion, and is sympathetic to his plight. In contrast, the Africans killed in the movie all die unlamented, without the character alignment afforded the chimpanzee. Similarly, whereas the pigmy bushmen who attack the whites are represented as cruel and savage, the elephants are presented as heroes when they attack the pigmy village and rescue the white captives. The movie, then, establishes hierarchies between human and animal, and European and African. The animals serve Tarzan because they recognize his mastery over them. However, they are also represented as more civilized than the savage African tribespeople.

The ambivalence towards wild animals remains at the end of the movie, when the adventurers finally arrive at the elephants' graveyard. 'It's beautiful. . . . solemn and beautiful', Jane declares, adding: 'We shouldn't be here'.

Harry Holt takes a more materialistic view: 'It's riches, millions'. In the closing moments of the film, Jane herself seems to accept this view, talking enthusiastically about their next big safari with Holt as its leader. 'Only this time there'll be no danger', she says, 'because we'll be there to protect you, every step of the way'. Her earlier recognition that they 'shouldn't be here' is thus disavowed in this final affirmation of imperialist permission. The closing image in the film shows Tarzan and Jane, with the chimpanzee in her arms, as they survey the landscape from a rock: the beneficence of the imperial project of mastery over Africa guaranteed by the maternal protection Jane offers the chimpanzee.[4]

African conservationist narratives

Hollywood's adaptations of two of Hemingway's African safari stories, filmed as *The Macomber Affair* (1947) and *The Snows of Kilimanjaro* (1952), revealed a crisis of confidence in the rugged masculinity associated with the great white hunter as the United States emerged from World War Two. *The Roots of Heaven* (1958), directed by John Huston from a script based on the Prix Goncourt-winning novel by Romain Gary, extended these doubts into a full-blown critique of big game hunting, and a pioneering call for the conservation of African elephants, placing the issue of wildlife conservation within a Cold War context of global crisis.[5]

Morel (Trevor Howard) is a disillusioned former big game hunter campaigning to stop elephant hunting. His concern for elephants began as a prisoner-of-war in Germany, when the image of elephants came to stand in his imagination for the freedom and space he had been denied by the Nazis:

> Anyone who has seen the great herds on the march across the last free spaces of the earth knows they're something the world can't afford to lose. But no, they have to capture, kill, destroy everything. All that's beautiful has to go. All that's free. Soon we'll be alone on this earth with nothing left to destroy but ourselves.

The wild animal has thus come to signify the freedom of the individual in a totalitarian society. Nevertheless, Morel's stand against both the killing of elephants for the ivory trade and the collection of specimens for zoos is based on biocentric rather than anthropocentric motives. His criterion for conservation is the suffering of individual animals, which, he tells the barmaid Minna (Juliette Greco), suffer for days when captured in traps. Minna is the first person to sign his petition, because, as a victim of sexual abuse during the war, she empathizes with the suffering of the animals. The movie constructs Morel's concern for animal welfare as morally superior to the more anthropocentric motive for saving the elephants put forward by the second man to sign up to help him, the drunken English military man Forsythe (Errol Flynn). 'This man Morel isn't really defending elephants', Forsythe tells Minna, 'he's defending us. We're all threatened with extinction. One more of those Hydrogen bombs, a couple of those Sputniks whizzing around up here, and we'll all go up in smoke like the poor ruddy old elephants'. Yet Minna's reply is closer to Morel's true motives: 'He might just happen to be fond of elephants'.

The Roots of Heaven identifies concern for African wildlife as a virtue solely confined to white Europeans. For the indigenous Africans, in contrast, the elephants are obstacles to the modernization of their country. 'How can you build a modern country with things like that in your way?', comments the African driver when his truck is held up by an elephant. Waitari, leader of the Free Africa Movement, considers the elephants as important only in so far as they can be exploited for the cause of African nationalism. Elephants are the symbol of African freedom, he tells Morel: 'They will be the emblems we will put on our flags and when we have the power firmly in our hands, we will protect them'. However, Waitari is represented negatively in the film as a cynical politician who gives his support to the poacher Habib, and plots to have Morel killed so that he can be presented as a martyr for the African nationalist cause.

The movie is thus careful to identify conservationism as a liberal humanist cause that is above the political manoeuvrings of Waitari, as Morel's conservationist manifesto for the World Committee for the Protection of Nature makes clear:

> The Committee recalls that it has no political character and that consideration of ideology, doctrine, Party, race, class and nationality are completely foreign to it. . . . It appeals solely to the feelings and dignity in every human being without discrimination and with no other thought than to call for a new international agreement on the protection of nature.

Morel's campaign for the humanistic treatment of wild animals is part of his belief in evolutionary progress. For Catholic priest Saint Denis, however, killing is an unchangeable part of fallen human nature. The killings 'will go on', he says, 'whatever the conference decides. We have it in our blood'. A man in the bar had earlier produced a similar argument. 'A man has got to eat, to hunt to kill, and sometimes be killed himself', he tells Morel. 'That's a fact of nature'. Morel replies: 'It's a fact of nature that we should change'.

The ending of the film, like that of Romain Gary's novel, leaves this question unresolved. Morel becomes disillusioned with his fight, and, in a moment of self-doubt, believes that the 'Habibs of the world are right', and that killing may be an unalterable fact of human nature. But Minna exhorts him to carry on with his struggle. When the French colonial troops allow him to escape with what remains of his followers, he walks off into the distance, temporarily defeated but alive to carry on fighting for his cause.

Although made a few years later than *The Roots of Heaven*, *Hatari!* (1962), the story of a band of Euro-American animal catchers hired by a Swiss zoo, is more equivocal in its attitude to the African safari. Indeed, that the film-makers did not possess a conservationist sensibility may be seen by the fate of the animals used in the production of the movie. After filming, the animals were not set free, but were given to the Tanganyikan government and to zoos in the United States.[6]

In keeping with the movie's title (the Swahili word for 'danger'), director Howard Hawks filmed the encounters with wild animals from the perspective of the hunters in a way that evokes the thrills of the chase. The use of location shooting, long takes, fast tracking shots taken from camera cars, telephoto close ups and multiple camera units, all add to the sense of kinetic thrill. The actors extemporized with the animals, and were not allowed stunt doubles. At various points, a buffalo and a rhinoceros, clearly enraged, can be seen ramming the truck. In this way, the movie transposes the preoccupation of the western with male heroism, physical prowess, leadership and teamwork, to the new frontier of Africa, which becomes a testing ground for American manhood, as nature poses a series of challenges to male ingenuity and physical bravery. In the most elaborate attempt to master nature in the film, Pockets (Red Buttons) successfully uses a rocket to fire a net over a tree to capture five hundred monkeys at once.

Yet the movie's attitude to such acts of masculine conquest is ambiguous. The men know that their work is dishonourable, observes critic Jean Douchet, and the 'resulting prickings of conscience explain why they are so prompt to yield to nature, to offer it a love that flies in the face of their man's nature'.[7] The nature to which the men are attracted, typically for a Hawks movie, is identified with the female. Dallas (Elsa Martinelli), the greenhorn photo-journalist hired to take pictures of the animal captures, shows a concern for, and affinity with, wild animals that is in marked contrast with the men's attitude towards them. 'I like animals', Pockets admits to her, 'I'm just scared of them'.

When a British game ranger kills a 'rogue' mother elephant because she attacked his family, Dallas saves the orphaned calf from being shot by Sean (John Wayne), and then weans it with a bottle. Pocket later tells her that Sean, despite his appearance to the contrary, 'didn't want to see it shot anymore than you did'. Dallas is naturally maternal towards the animal, whom she calls Baby. When she dresses up as Mother of the Elephant in a tribal ceremony, the white woman is further positioned as nurturer and protector of nature, a crosser of boundaries between white and black, human and animal.

When Dallas finally abandons hope of attracting Sean, he chases after her with the elephant Tembo on her scent. The woman is thus animalized, and becomes the male hunter's ultimate catch. In the final scene, the couple's marriage bed collapses under the weight of an invasion of baby elephants. Nature, in the guise of animals and women, thus comically disrupts masculine attempts at order and control, and is consequently both admired and denigrated. *Hatari!*, then, provides a complex, nuanced version of the safari narrative, while avoiding the open advocacy of wildlife conservation pioneered by *The Roots of Heaven*.

In the 1970s, producer-director Noel Marshall and actress Tippi Hedren emerged as animal welfare advocates, founding the Shambala Ranch in California for the care of wild animals used in Hollywood film production, and extending their concern for endangered species to the making of conservationist advocacy movies. *Mr. Kingstreet's War* (1973) is set in East Africa at the start of World War Two, and laments the loss of 'game' species due to human greed and stupidity.

American Jim Kingstreet (John Saxon), a former hunter turned conservationist, tries to protect African wildlife threatened by the Italian and British armies, who both want to use a local waterhole as a base for their operations. The waterholes, he tells the British Colonel, 'supply an African village, they keep a lot of game from dying of thirst'. If he does not protect them, the consequences will be dire: 'the game shot at or driven away, every animal we've coaxed back here driven to extinction'.

In a similar way to *The Roots of Heaven*, Kingstreet's neutrality marks conservationism as apolitical and above national rivalries. It is instead viewed as the product of American frontier idealism, in which the white male hero is supported by his dedicated wife, Mary (Tippi Hedren), and his loyal African servant, Jomo. Although Jomo is given a somewhat larger role than is common in Hollywood movies about Africa (he even suggests to Kingstreet the idea of using grenades to sabotage the British fuel dump), leadership nevertheless belongs to the white American male, with Africans as subordinates. This American paternalism is further reinforced by Kingstreet's relationship to the local tribespeople, who were saved by his windmill after their well had dried up. The village chief is grateful and deferential to Kingstreet. 'You are our friend, you are father, you are my grandfather', he tells him. The Chief therefore accedes to Kingstreet's request to stop local youth using snares to trap wild animals. Significantly, then, the movie criticizes traditional African hunting practices, while at no time questioning the American hero's seemingly natural ownership of land in Africa.

In an unexpected denouement, however, *Mr. Kingstreet's War* ends with the failure of the American hero. Jim's evil, mad brother, Morgan (Brian O'Shaughnessy), an unrepentant ivory poacher, is also his bad conscience and nemesis. In a childhood accident, recounted in flashback, Jim accidentally poked his brother's eye out with a pair of scissors. Consumed by bitterness,

Morgan becomes Jim's 'cross', telling him: 'You'll carry me 'til one of us is dead'. Morgan brings about the failure of Jim's conservationist mission, when, after being beaten up by Italian troops, he seeks revenge on Major Bernadelli (Rossano Brazzi), the Italian fascist leader. In the final scene of the movie, Bernadelli walks towards Kingstreet, extending his hand in a gesture of apparent conciliation, when he is shot dead by Morgan. In the ensuing gun battle, all the major characters are killed in an excessive, apocalyptic, post-*Wild Bunch* (1969) orgy of slow motion dying and distorted synthesized sound. The non-conformist American hero has failed, this time, to halt the forces of history. His demon brother and repressed shadow self have destroyed his conservationist dream. The movie thus confirms the pessimistic view of Major Bernadelli at the start of the film. 'You cannot change the world', he tells Kingstreet. 'No-one can do it'.

The romantic myth of the white hunter, melodramatically demolished at the end of *Mr. Kingstreet's War*, has also been maintained in movies such as the Oscar-winning *Out of Africa* (1985), in which Robert Redford plays the figure nostalgically as a charismatic, rugged individualist. In contrast, *White Hunter, Black Heart* (1990), a fictionalized account of director John Huston's quest to shoot an elephant before he began filming *The African Queen* (1951), uses an equally charismatic actor, but is more concerned with questioning and under-mining the heroism of its central protagonist, John Wilson (Clint Eastwood).

At the start of the film, Wilson is presented as an admirable, inner-directed hero: intuitive, instinctive and physically brave. His anti-racist credentials and desire to protect the vulnerable are shown when he insults an anti-Semitic British woman, and fist fights the general manager of the hotel who had kicked a black servant for spilling a drink. Wilson also shows himself to be a man of integrity, refusing to sacrifice his artistic ideas to commercial imperatives and audience tastes. However, when he delays the production of his new movie to pursue his desire to hunt an elephant, Wilson proves himself to be arrogant and careless, his rugged individualism recast as selfishness and a lack of consideration for the needs of the team. Eventually, his recklessness leads to the death of Kivu, his chief tracker, whose skills he had respected.

As the figure of Wilson is opened up to criticism, so the views of the scriptwriter Pete, who believes that killing elephants is wrong, are vindicated. Wilson's passion for hunting, according to Pete, is 'just like any passion: it's irresponsible and it's destructive'. When Wilson accuses him of 'playing it safe, because you're scared and you know it', Pete replies, 'Alright, well, I guess I'll just have to live with that'. Pete is the modern man of compromise, whose values ultimately gain ascendancy in the movie.

Pete has a greater respect for the elephants than Wilson, and does not wish to kill them. When he sees elephants in the bush for the first time, he accepts their sublimity, as he comments to his friend Hod:

> So majestic, so indestructible. They're part of the earth. Make us feel like perverse little creatures from another planet, without any dignity. Makes one believe in God, in the miracle of creation. Fantastic. They're part of a world that no longer exists, Hod. A feeling of unconquerable time.

In the end, Pete's respect for the elephants influences Wilson, who is unable to shoot an elephant when finally given the opportunity to do so.

Despite this undermining of the myth of the heroic white hunter, however, the representation of Africans in *White Hunter, Black Heart* remains problematic. That the hunting party returns to the village without Kivu's body is not explained. Instead, the ending dwells not on the African victim but on the guilt of the white man, as black self-sacrifice makes Wilson weep.

As the reputation of the white hunter has declined in Hollywood cinema, so the reputation of African wild animals has been rehabilitated. The remainder of this chapter will examine the rise of conservationist advocacy with regard to two charismatic species of Africa fauna: the lion and the mountain gorilla.

The changing image of the African lion

Safari movies such as *Trader Horn* constructed the African lion as a savage beast and obstacle to imperial progress. In the pioneering 3D movie *Bwana Devil* (1952), savage lions not only hold up the building of a railroad, but appear to attack the spectator. *The Ghost and the Darkness* (1996), reworking the same historical material as *Bwana Devil*, returned to its projection of hyperbolic savagery onto the animal. In both films, the eventual killing of the savage lions by the white male hero allows for the continuation of British imperial interests in Africa.[8]

In the 1960s, however, a counter-tendency to such representations began to develop. A transitional movie in the cultural shift towards the image of the good lion was the Ivan Tors production *Clarence the Cross-Eyed Lion* (1965), which made the African lion not only tame, but ridiculous, humanizing the animal for comic purposes in the tradition of the circus. The villains of the movie are poachers, and the hero a widowed animal doctor on a game reserve, with a paternalistic respect for nature. Dr Tracy's daughter Paula keeps Clarence as a pet, because he is too cross-eyed to hunt for himself. That male lions do not hunt anyway is a fact of natural history in which the movie is uninterested, suggesting both its lack of desire for authenticity and the conservative gender assumptions which it ultimately endorses.

Clarence, more lamb than lion, is lightly disruptive of order and decorum: belching and knocking over household objects. Much of the comedy is at the expense of the effete, upper-class English tutor, Rupert Rowbotham, who is terrified of wild animals, and advises Paula that 'no, absolutely no animal is harmless', and that she should treat animals with 'sensible and proper suspicion, and avoid all personal contact with them whatsoever'. Clarence's behaviour confirms the opposite of Rowbotham's paranoiac and ignorant view of animals.

A more positive attitude to wild animals is shown in the film by anthropologist Julie Harper, a Dian Fossey figure studying mountain gorillas in the forest. The plot of the movie concerns the threat to Julie's work posed by rebel African terrorists, led by a French poacher, who plan to capture gorillas from the rain forest to trade for weapons. When warned by Tracy of the danger she is in, Julie refuses to leave the forest, saying she is happy in her work. 'I care about them', she says of the gorillas. 'They're my friends'.

Predictably, when Tracy is captured trying to save Julie from the rebels, Clarence comes to the rescue by attacking the terrorist leader. Julie then admits to Tracy that she 'should have left when you told me to'. Accepting his proposal of marriage, she tells him it feels like coming in 'out of the rain'. The outdoor adventure movie thus again constructs gender relationships in a conservative way: the independent working woman wants marriage above all else, her work less important than her role as love interest for the widowed doctor. In the African wilderness, the benevolent paternalism of the American father is thereby reaffirmed.

As a proto-conservationist narrative, *Clarence the Cross-Eyed Lion* demonstrates an interest in wild animal welfare, but not yet in animal release. At the end of the movie, Tracy gives Clarence a pair of corrective spectacles, and Paula tells him: 'Now that daddy's fixed your eyes, you can stay with us forever and ever. Isn't it more comfy in here, than out there in the bush with no-one to play with?' So the girl, like Sandy in Ivan Tor's *Flipper*, made three years before, keeps the tame animal as a pet, for the entertainment and therapeutic value he holds for his human owner.

In contrast, the Anglo-American production *Born Free* (1966) indicated a new conservationist sensibility in popular cinema, by endorsing the release of lions from captivity. Nevertheless, the movie still displays a neo-colonialist attitude by limiting African characters to the marginal role of servants, stereotypically docile and faithful, and unimportant in terms of either characterization or plot. Moreover, in the scenes shot in the bush with the lioness Elsa, local tribespeople are entirely absent. In this way, the movie denies African people agency in relation to their own environment, and instead idealizes the benevolent paternalism of its two white British conservationists.

Born Free begins with a lion stalking and killing an African woman washing clothes in a river. Game warden George Adamson (Bill Travers) shoots the man-eater dead, but is then attacked by his mate. He kills the lioness in self-defence, and then discovers that she was protecting her three cubs, now orphaned. The ensuing narrative centres on the relationship between Joy Adamson (Virginia McKenna) and the three lion cubs, for whom she becomes a surrogate mother. When the cubs have grown too big to keep, the District Commissioner orders Joy to donate them to Rotterdam Zoo. But Joy refuses to give up her favourite, Elsa. The Adamsons' subsequent decision to return Elsa to the wild is the defining moment of the conservationist narrative, and is the crucial difference between *Born Free* and a transitional movie like *Clarence the Cross-Eyed Lion*. *Born Free* endorses the active intervention of human beings in the processes of nature to preserve the life of individual animals in the wild, within an explicit discourse of animal rights. Joy justifies her refusal to allow Elsa to be sent to the zoo by saying that the lioness 'was born free, and she has the right to live free'.

The 'freedom' celebrated by *Born Free* is also, of course, a fantasy of human freedom projected onto the wild animal. When George says that Elsa will be 'safe in the zoo', Joy replies: 'Yes, safe, and fat and lazy and dull and stupid like some cow in a milking machine'. Elsa, then, stands for individual freedom denied by industrial society, and rediscovered in a rural authenticity superior to the mechanistic regime imposed not only on cows, but on city-dwelling organization men. 'Why don't we live in some nice, comfortable city, George?', Joy asks. 'Other people do. But we've chosen to live out here because it represents freedom for us, because we can breathe here'. As in *The Roots of Heaven*, then, the wildness of the animal connotes a notion of liberal freedom threatened by modernity.

However, this construction of the lioness Elsa as a desirable form of wildness involved the suppression, from both the book and the movie, of signs of excessive or unacceptable aggression. Biographer Adrian House records two incidents excluded from the book *Born Free* on the insistence of Lord William Percy, whom Joy Adamson had engaged to write the preface: one in which Elsa bit the arm of a white hunter, and a later incident in which a male lion introduced as a potential mate for Elsa attacked several local people before being killed by them. Similarly, the movie represents Elsa as comically playful and mischievous, but never overtly aggressive towards human beings, even though the shooting of the film resulted in several injuries to the cast from the lions used in the production.[9] Though not sentimentalized and anthropomorphized to anywhere near the extent of Clarence the Cross-Eyed Lion, Elsa is nevertheless sanitized in order to fit the dual ideological project of the movie: that of endorsing animal reintroductions, and of constructing a desirable symbol of freedom for a Cold War society.[10]

The symbolic association of the lion with spontaneity and freedom is, of course, the effect of particular cinematic techniques. For producer Carl Foreman, the lack of special effects in the movie guaranteed its authenticity.

The film-makers, he said, rejected 'any faking or phoniness or "double exposures". We would either find another lioness that could do what Elsa did—and thus prove Mrs. Adamson a true prophet—or we would go down with flying colors'. Following advice from George Adamson, Foreman replaced the circus lions originally intended for the film with untrained lions bred in Kenya, and affection-trained by the cast with the help of George Adamson. Pre-trained or tame lions would, as Foreman put it later, be 'stolid, static, conditioned in the wrong way—either to fear of blows or fear of inescapable bars'. As for Elsa, however:

> Spontaneity of behavior was her chief attribute. We had to find a free-born animal, like Elsa, and with no 'complexes' already established. That meant, of course, real danger, for while Elsa didn't turn on Mrs. Adamson, was there any guarantee that the film prototype of Elsa would remember the demands of her script and behave accordingly?[11]

The contradiction between an animal behaving spontaneously and remembering her script is a reminder of a fundamental aspect of Hollywood wild animal movies: the need for the film-makers to fake nature in order to simulate authenticity. In *Born Free*, the lions were both trained by George Adamson and constructed cinematically to appear wild and spontaneous. The pursuit of authenticity involved allowing the unpredictability of the lions' behaviour to produce extemporizations by the actors. 'I may have been the director', commented James Hill, 'but the lions told us all what to do'.[12] The genuine sense of risk and danger in using semi-trained lions added dramatic tension to the actors' performances.

The desire for authenticity in *Born Free* was a self-conscious attempt to differentiate its representation of wild animals from the humanizing circus tradition still evident in movies such as *Clarence the Cross-Eyed Lion.* Like Clarence, the lion cubs in *Born Free* are comically disruptive of domestic order, urinating on the floor and stealing George's towel while he is washing, but the movie avoids making them too cute and gentle, and maintains a tension between friendliness and ferocity, terror and affection, the sublime and the beautiful. Elsa, a lion re-trained by human beings to be wild, and taught how to be natural, is ambiguously positioned between wildness and habituation, and thereby mediates the alienated relationship between human beings and nature. In the end, Joy Adamson too has the best of both worlds, as George's words to her are fulfilled: 'you're hoping. . . . she'll be wild but not too wild. That you can see her every now and again'.

While the ambiguity of Elsa's position between wildness and civilization is a source of *Born Free*'s abiding power as fable, it is also the basis of the criticism that George Adamson's approach to lion conservation has attracted.[13] On the completion of *Born Free*, Columbia Pictures had finally confirmed the

Adamsons' suspicions as to the limits of its concern for animal welfare, by planning to sell all of the lions used in the production to zoos and safari parks. George Adamson was able to save only three out of the twenty-four lions used in the film from this fate, and established the lion release project in the Meru reserve to prepare them for future release into the wild.[14]

However, George's policy of continuing to feed and keep in touch with the lions after he had returned them to the wild, an action endorsed by the end of *Born Free* itself, was considered dangerous by several interested parties. Local African hunters and herdsmen, forbidden by law to practice their traditional ways of life in the reserved areas, opposed his scheme, as did expatriate safari guides and professional game managers. All agreed that a once-tamed lion, being too familiar with people, is potentially dangerous when returned to the wild. Joy Adamson herself also opposed her husband's lion release project on these grounds. Such fears appeared to have been borne out when, in 1971, the released lion Boy was shot dead by George Adamson after mauling to death an African worker at the camp. The animal had found it difficult to fend for himself in the wild, and had already been gored by a buffalo and attacked by two wild lions.[15]

In defence of George Adamson's methods, biographer Adrian House asserts that the lions he released 'were no more dangerous than wild ones—almost certainly they were less so because the human beings they encountered were friendly, not hostile or provocative'.[16] Moreover, House found no evidence that George's lions had attacked the local population. Those attacked, he pointed out, were workers who had disobeyed strict instructions regarding their behaviour towards the lions.

Born Free became a model for the wild animal release movie, not only in its narrative structure, but also in the pioneering techniques brought to the filming of animals by director James Hill, the Adamsons and the actors Virginia McKenna and Bill Travers.[17] The improvisatory techniques used in the film were developed even further in Noel Marshall and Tippi Hedren's main conservationist project, *Roar* (1981), which was directed by Marshall himself, and featured himself, Hedren and their children in acting roles. Although shooting began on Marshall and Hedren's Shambala Ranch, in Soledad, California, in 1969, the movie was not completed until 1981, with production delayed by flood, fire, disease and the many serious injuries to cast and crew inflicted by their contact with the wild animals with which they were working.[18]

The plot concerns zoologist Hank (Noel Marshall), who has left his wife and children to spend three years in an animal foundation in Africa. Hank is a Wild Man figure, appropriately hirsute, who believes in getting close to the lions, even to the point of physically breaking up their fights. His stated ambition is to make a comparative study of the great cats of Africa and 'to become almost a member of the pride'. The gender implications of such a desire play a central role in the film.

At the start of the film, Hank's wife Madeline (Tippi Hedren) and children travel to Africa to be reunited with their father. Madeline reassures her daughter Melanie (Melanie Griffith) that her father left home because of his work, and 'not because he didn't love us'. The sense of disharmony in the family is compounded when Melanie accuses her mother of being the cause of the family's separation.

When Madeline and the children arrive unexpectedly while Hank is away, they are subjected to a frightening ordeal, as lions and tigers range freely through the house. The *mise-en-scène* constructs a documentary authenticity through hand-held close-ups and long takes, interspersed with occasional quick cutting. Six cameras were used to film these sequences, and Marshall refused to let his actors use stunt doubles, increasing the sense of fear and danger. At one point, Melanie appears to become hysterical. The opening credits noted, somewhat preciously: 'Since the choice was made to use untrained animals and since for the most part they chose to do as they wished, it's only fair they share the writing and directing credit'.

After this extended sequence of vicarious, circus-like thrills, *Roar* ends in bathos. Having finally escaped from the wild animals, the family awake next morning to find the lions and tigers asleep next to them. The father arrives, belatedly, to find all the animals friendly. Implausibly, the animals' behaviour has radically changed. Hank then explains away this near-miraculous transformation as the product of a simple misunderstanding: his family had simply misread the animals' behaviour as aggressive, when they were only trying to be friendly.

That the lions turn out to be gentle rather than aggressive serves the ideological project of the movie, which is to reaffirm the authority of the human male as head of his newly harmonious family. Hank had earlier established a parallel between human beings and lions, explaining about the dominant lion, Robbie: 'Their whole life is based on dominance. That's why we have such a great harmony here, 'cos Robbie is such a gentle and loving ruler. He also knows how to protect the others from outsiders'. The lions thus provide an ideal of natural family order that is later enacted by the human beings in the narrative.

That the lions are friendly towards human beings also serves the conservationist message of the movie, as stated explicitly by Hank: 'We can't keep exterminating them. We can't keep eliminating their land. We can't keep exterminating everything that we fear, or that inconveniences us'. Here, the vague use of the third person plural ('we') elides issues of agency and responsibility, and the structures of power they imply. Agency is instead casually attributed to a vague, undifferentiated human nature, rather than specified in terms of relevant social factors, such as, for example, nationality, race, gender or class. The movie's lengthy postscript hammers home its conservationist message, while further obfuscating the issue of agency, this time by using the passive tense: 'In the eleven years since we began filming *Roar*, in most areas of Africa, 90% of the animals *have been killed*. These thinking, feeling beings who need your help to survive.' (italics added). The passive tense avoids the issue of *which* human beings, within *which* social structures, have killed the animals. *Roar*, then, ultimately obscures the question of agency in both environmental destruction and conservation, thereby emptying conservationism of politics.

Gorillas in the Mist and the conservation of mountain gorillas

The environmentalist politics of *Gorillas in the Mist* (1988) are shaped by the conventions of the Hollywood biopic genre.[19] Starring Sigourney Weaver as the primatologist Dian Fossey, who studied mountain gorillas for thirteen years in East Africa until her unsolved murder in 1985, the movie celebrates Fossey as a heroine of conservationism, in a narrative of heroic self-realization and noble sacrifice. It achieves this by being highly selective of the biographical and historical evidence available on Fossey's life and work. Indeed, journalist Nina J. Easton records that director Michael Apted decided to refrain from talking to Fossey's former associates because of the negative impressions they were giving of her. 'The high-wire act of the film', he told the *Los Angeles Times*, 'was to see how tough-minded I could make her without losing the audience's sympathy for her'.[20]

The film particularly omits the criticisms that Fossey's conservationist actions attracted from the international wildlife conservation movement, because of her poor relationship with the local African population, an aspect of her work that the movie also glosses over.[21] Instead, it draws on highly selective examples of Fossey's campaign of direct action (or 'active conservation', as she called it) waged against local poachers and farmers who were threatening the existence of the mountain gorillas. She is shown threatening to hang one of the poachers, burning down their huts, and posing as a witch in a Halloween mask to scare a young boy. Through character align-ment and allegiance, the movie justifies these actions as expressions of both her passionate character and her deep love for the gorillas. She is an American vigilante hero, willing to break official law to uphold her inner sense of justice.

In order to encourage audience sympathy for Fossey in this way, the film-makers omitted several incidents reported by her biographer Harold Hayes which may have presented her in a more negative light. For example, her actions towards native Rwandans also included whipping the testicles of a captured poacher with nettles, pistol-whipping another, stuffing sleeping pills down their throats, and shooting thirty head of cattle as a warning to farmers. The movie thus sanitizes the violence used by Fossey, in order to maintain her heroic status.[22]

Fossey's aggressive attitude towards the local African population is, then, largely vindicated in the film, despite a short scene which complicates its representation of the politics of conservation by suggesting that it is white Americans who are ultimately responsible for the killing of the mountain gorillas. When Fossey refers to the local tribe as the 'goddamned' Batwa, *National Geographic* photographer Bob Campbell (Bryan Brown) corrects her. 'You can't put all the blame on the Batwa', he tells her. 'They've been feeding their families like this for generations. If you want to blame anyone, blame the doctor in Miami. He's the one that hires the bloke that hires the Batwas. The Batwas get to feed their kids, the middle man gets a silk shirt, the doctor gets a gorilla hand ashtray for his coffee table'. When Fossey answers that she 'can't get to the damn doctor in Miami', Campbell suggests that working for *National Geographic* will help to publicize gorilla conservation in the United States.

Significantly, the movie does not follow through the suggestion made by Campbell that demand for gorilla trophies by consumers in the United States is the main cause of poaching in the Virungas. Instead, it concentrates on the heroic efforts of its American protagonist to overcome the obstacles put in the way of gorilla conservation by the local African population. Thus what may be interpreted as Fossey's high-handedness and contempt for local African politics ('This place is a disaster, some little civil war', she says as she is evicted from the Congo) can be justified as a sign of her humanitarian concern for the gorillas, which nobly transcends politics. Gorillas, she observes, 'don't know borders. They don't need passports'. Similarly, her later meetings with the Rwandan government official, who seeks to justify selling gorillas to zoos on economic grounds, can also be interpreted as evidence of her passionate dedication to gorilla preservation. When the official tells her that protection of park lands is expensive, she angrily replies: 'That's your problem: make new laws, raise taxes, but give my gorillas the protection they're entitled to'. Fossey's attitude may be interpreted as arrogant and colonialist, but it is also motivated by righteous indignation and a desire for justice for the gorillas.

Moreover, the movie is careful to disavow any suggestion of colonialism or racism in Fossey's attitudes to the local African population by inventing the character of Sembagare as her devoted African guide, apparently in love with her. The final image of Sembagare is of the trusted subaltern, faithfully tending Fossey's grave. This special pleading for Fossey's benevolence perpetuates the typical colonialist representational strategy in which white Euro-Americans act on behalf of a passive and grateful Third World populace.

As mentioned earlier, *Gorillas in the Mist* also idealizes Fossey's role in modern conservationism by omitting to mention the opposition that her methods aroused within international wildlife conservation groups. Official conservation policy supported the establishment of the Mountain Gorilla Project, which aimed to preserve the gorillas and their habitat by developing tourism, thereby appealing to the economic benefits that gorilla preservation would have for local people. Although she did not rule out all long-term conservation projects, Fossey opposed this project because it appeared to her to be too slow.[23] The movie conveys Fossey's opposition to tourism when Sembagare mentions her attempts to discourage tourists by spreading rumours of typhoid in the mountains, and by firing shots over their heads. Fossey replies adamantly: 'They are not gonna turn this mountain into a goddamn zoo'. Again, her vigilante actions are justified as an expression of her righteous indignation and passionate determination to preserve the gorillas.

What is not mentioned in the movie, however, is that her deteriorating relationship with the local African population led to her eventual replacement as head of the Karisoke Research Center and to the ending of funding for her conservation activities. She was also blackmailed by poachers, who threatened raids on her study groups if she did not look after sick gorilla infants destined for zoos. When a gorilla infant died in her care, poachers retaliated by killing more gorillas. At the time of her murder, probably by poachers, international groups were calling for her to leave Rwanda.

The movie ignores these details, concentrating instead on the conventional conflict in a Hollywood 'woman's film' between the female protagonist's career ambitions and her love relationships. Although Fossey swears and chain-smokes (Hollywood code for a strong-willed and independent woman), the movie also represents her as traditionally feminine enough to throw a tantrum at the airport upon arrival, when her cases, including her hairdryer, make-up, underwear and brassières, have to be left behind before her trek into the rain forest. The movie's construction of Fossey as a figure of feminist independence is thus cautious and contradictory. Nevertheless, the film explicitly celebrates Fossey's decision to choose her calling over conformity to the traditional expectations of her gender. Faced with the thwarting of her ambitions to work with the gorillas after being thrown out of Congo during the civil war, Fossey is shown overcoming social pressure to conform to traditional notions of female behaviour. She tells her friends: 'I'm going to go home, I'm going to buy the sexiest dress I can find, I'm going to marry David and he's never going to hear another peep out of me'. However, she instead rejects marriage for her career. Although the outcome of this choice is shown to be isolation and loneliness, it is also presented as a noble sacrifice taken for the gorillas she loves.

Later, when Bob Campbell suggests that they leave Africa for half a year at a time, Fossey ends the relationship. The scene of the lovers' parting is followed by a shot of Fossey alone with the gorillas in the rain. The movie then cuts to five years later, with Fossey noticeably more withdrawn from society and suffering from emphysema. At this point, the movie may seem to be endorsing the commonplace patriarchal moral that women cannot find fulfilment in both a career and their emotional lives. Indeed, Vera Norwood sees the movie as a veiled warning against women's liberation, in its apparent implication that

> the arc of Fossey's life began its downward slide into violence and death when she rejected a marriage proposal from a National Geographic photographer. Renouncing traditional bonds of home and family, she placed herself in jeopardy. . . . her allegiance to wild beasts made her a wild woman, leaving her vulnerable to the violence visited upon other animals. Fossey, popular culture suggests, was killed once she stepped outside the bounds of domestic space and into the landscape of wilderness.[24]

However, Norwood's judgement on Fossey ironically confirms the negative judgement passed on her in the movie itself by the villainous animal collector Van Vecten: 'Crazy woman. You go too far'. Yet the movie's representation of Fossey can be defended on realist grounds, as accurate to its biographical sources. Norwood's interpretation does not take into account the evidence from Fossey's friends and biographers, and from her own diaries, that her broken relationship with Bob Campbell *did* have a profound affect on her mental state. 'Never have I known such sorrow', Fossey herself wrote in her diary on the occasion of Campbell's departure. Her friend Alan Root confirms that she was emotionally shattered when Campbell left.[25]

Moreover, the movie does not simply portray Fossey's life as a 'downwards spiral', as argued by Norwood. As already noted, her self-sacrifice is represented as a noble one. When she castigates a research scientist for his 'me-itis', and flies into a rage at finding two students in bed together, her behaviour is again justified by her love for the gorillas, and her frustration at their treatment at the hands of the poachers. At the emotional climax of the film, Fossey sobs over the killing of Digit, her favourite gorilla, crying in despair: 'They took his head!' Rather than criticizing Fossey, then, the movie celebrates her life and achievement, and presents her murder as unjust, not as a punishment for her supposed transgression.

As for the representation of the mountain gorilla in the movie, *Gorillas in the Mist* successfully revises previous cinematic images of the animal by relying for its sense of authenticity on wild gorillas, rather than gorillas habituated to tourists. Occasional scenes, such as the killing of Digit by poachers, used actors in gorilla suits, but the relative lack of such simulations gives a sense of authenticity lacking in a similar movie about benign gorillas and animal rights, *Greystoke: The Legend of Tarzan, Lord of the Apes* (1984). Nevertheless, compared to the gorillas in Fossey's book, who practice infanticide, masturbation, incest, fellatio and cannibalism, the animals in the movie are idealized figures possessing the redemptive innocence typical of the Hollywood wild animal movie.[26]

Many of the movies discussed in this chapter have helped to publicize wildlife conservation as a global issue, and successfully raised money for those conservationist organizations they advocate, such as the Elsa Wild Animal Appeal (later the Elsa Conservation Trust) in *Born Free*, and the Digit Fund and the World Wide Fund for Nature in *Gorillas in the Mist*.[27] Nevertheless, the preservationist practices advocated by both the Adamsons and Dian Fossey have been criticized, by both scientific conservationists and social ecologists, as having led to the forcible expulsion of local African people from their traditional lands, and the abandonment of traditional subsistence hunting practices that have been carried on for centuries. Hunters have been redefined as poachers, without full consideration of the social and political contexts of species extinction. As conservationists Adams and McShane put it: 'as long as conservation operates on the notion that saving wild animals means keeping them as far away as possible from human beings, it will become less and less relevant to modern Africans. Parks and other protected areas will eventually be overrun by people's need for land unless the parks serve, or are at least not completely inimical to, the needs of the local population'.[28] In environmentalist terms, then, Hollywood's African conservationist movies have focused on heroic westerners such as Dian Fossey and the Adamsons, and have accordingly tended to marginalize or over-simplify the role of African people in ecological issues, while evading complex political issues of social justice within the various ecosystems in which the movies are set.

III

DEVELOPMENT AND THE POLITICS OF LAND USE

Introduction

The American cult of wildness in many of the movies explored in Parts One and Two of this book is a historic product of industrialization and urbanization, offering a therapeutic retreat from the anxieties aroused by those forces. Yet Hollywood cinema has also produced topical narratives which investigate the effects that the processes of modernity have had on the land and its peoples. In the movies discussed in Part Three, land is represented as both property and resource, a political space produced by intersecting forces of capital, labour and technology. Topical issues, as varied as the OPEC oil crisis, the Exxon Valdez disaster, the US farming crisis of the 1980s, and the death of anti-nuclear activist Karen Silkwood, provided starting points for narratives which drew on the aesthetic conventions of realism and melodrama to dramatize, and also, to a greater or lesser degree, displace, elide and mystify the power relationships involved in struggles over land use in the United States.

For example, although farming narratives such as *The River* (1984) and *The Milagro Beanfield War* (1988) at times evoke the agrarian landscape as a rural idyll, their narratives crucially turn on power struggles over land, and the threat of eviction by big businessmen. They are, in Raymond Williams' terms, 'counter-pastorals' concerned with economic crisis and loss, and celebrating both individual and collective resistance to monopolistic power.[1] In these films, and many others discussed in this section, the melodramatic mode typically constructs ecological-political issues as a Manichean struggle between a demonized big business and populist heroes. It is in the strength and resistance of the latter, ordinary men and women of goodwill and moral innocence, that solutions are usually seen to lie.

Yet many of the movies discussed in Part Three avoid a firm sense of closure. Although Steven Seagal announces his faith in the efficacy of 'green' technologies, *On Deadly Ground*, like *The China Syndrome* before it, ends with only a temporary victory for its heroes, and a recognition that the larger struggle against the destructive forces of greed and capitalist expediency has not been won. Moreover, in *Chinatown* (1974) and *The Two Jakes* (1990), both scripted by Robert Towne, the development of Los Angeles involves power hierarchies too entrenched, and human corruption too fundamental, for isolated individuals to challenge them. The conspiracy thriller *The Formula* (1980) ends in a similar impasse.

The range of films analysed in the following pages is, of course, necessarily selective: the representation in popular American cinema of such ecological concerns as food production and consumption, housing, mining, railroads and airplanes could all be explored within an ecocritical framework, but are beyond the scope of this book.

The analysis of land use and development in Part Three continues the critical approach taken throughout this book regarding human relationships with non-human nature. As such, it runs counter to two related tendencies that may be discerned in postmodernist and poststructuralist attitudes towards nature and environment of special relevance to the exploration of industrial and urban spaces.

The first is the idea of the 'death of nature'. Frederic Jameson writes:

> In modernism, some residual zones of 'nature' and 'being', of the old, the older, the archaic, still subsists; culture can still do something to that nature and work at transforming that 'referent'. Postmodernism is what you have when the modernization process is complete and nature is gone for good. It is a more fully human world than the older one, but one in which 'culture' has become a veritable 'second nature'.[2]

However, as environmental historian Donald Worster points out, in a response to Bill McKibben's *The End of Nature*, the argument that nature has been totally replaced by culture complacently assumes that the human 'conquest' of nature is complete, rather than an ongoing, provisional, two-way process that includes not only partial conquests of non-human nature on the part of human beings, but also adaptations to it as well.[3] The concept of 'adaptation', moreover, logically requires the very distinction between 'nature' and 'culture' that Jameson seeks to deny. 'One must question, in fact,' writes Kate Soper, 'whether there can be any claim to the effect that the nature–culture dichotomy is itself conventional, which does not tacitly rely for its force on precisely that objectively grounded distinction between what is humanly instituted and what is naturally ordained, which is being rhetorically denied'.[4]

The second critical tendency that the approach taken in this book challenges is the technophobia that sometimes comes to the fore when post-structuralist thought overlaps with deep ecology. In *Ecopolitics: The Environment in Poststructuralist Thought* (1997), Verena Andermatt Conley summarizes the contribution of structuralism and poststructuralism to the development of what she calls 'new ecological territories' as follows:

> Structuralism and poststructuralism were vital in separating language from referent and in showing how everything is a construction or a composition. They also evacuated subjectivity and opted for non-intervention in nature. A return less to an 'eco-subject' than to subjectivity is important as long as it integrates the lessons of these movements, in particular those of construction, non-mastery and of decentralizing humans.[5]

Yet Conley's conflation here of 'non-intervention in nature' with 'non-mastery' belies the confusion at the heart of such thinking. The real issue should not be whether human societies are to intervene in nature or not, but the exact conditions and extent of such interventions. 'Non-mastery' of nature may be a possibility for human societies, but 'non-intervention' in nature certainly is not.

The technophobia inherent in such deep ecological anxieties over human intervention in non-human nature finds its *tour de force* in the opening paragraph of Jhan Hochman's *Green Cultural Studies* (1998):

> The human obliteration of nature marks a twofold 'literation.' First, culture scrawls itself on nature's flesh. Animals are tagged, branded, genetically rewritten, fatally punctuated by bullets and arrows, and fatally scored by blades and traps. Animal skin is made into vellum and parchment. Trees, standing or pulped, are carved and written upon—their cellulose flesh processed into celluloid. Land is inscribed by rows, strips, grids, *bound*aries; rock is inscribed by explosion, cutting, painting, graffiti. Rivers are redrawn by dams, levees, and locks; and skies once thought to exhibit divine messages are *cursed* by airplanes leaving new messages in thin air, tainted (*tincta* means inked stroke) by exhaust from cars *cursing* the land below.[6]

Despite the powerfully emotive rhetoric, the totalizing position Hochman assumes leaves no place for effective resistance against environmental damage. For he denigrates all forms of human culture, including not only the modern technologies of automobile and aeroplane, but also agriculture and even writing. Ultimately, such a position is motivated by nostalgia for an enchanted, divinely immanent world, before the fall into human culture. Yet nowhere does Hochman admit the possibility of his own complicity in the evil forces of destruction, of which he writes with all the inverted, lurid, complicit relish of some anti-pornography campaigners.

A more nuanced, less all-or-nothing position regarding actual and possible relationships between human beings and non-human nature would give less ammunition to critics of radical environmentalism such as Gregg Easterbrook, whose description of what he calls the 'SOMEBODY ELSE logic' of some environmentalists should be carefully heeded. 'The idea that SOMEBODY ELSE should go without the conveniences of modern life', he writes, 'must be eradicated if environmental thought is to proceed to the next level of usefulness to society'.[7] It is not necessary to subscribe to the rest of Easterbrook's analysis of environmentalism to accept that he has here identified a problem with environmental discourses which rely on a rhetorical antipathy to human culture.

In Part Three of *Green Screen*, then, issues of development and land use are viewed as political struggles over appropriate ways for human beings to intervene in a non-human world conceptualized as a referent independent of those human discourses invented to explain it. The 'industrial-agricultural balance, in all its physical forms of town-and-country relations', writes Raymond Williams, 'is the product, however mediated, of a set of decisions about capital investment made by the minority which controls capital and which determines its use by calculations of profit'.[8] The complex politics of such relationships between human beings, and between human beings and non-human nature, cannot be reduced to a scenario in which 'development', 'technology' and even 'culture' are viewed merely as curses against an innocent nature.

TEN
Country and City

Cattle ranching and the western

The western genre has produced many stories of white homesteaders struggling to protect their farms from an impersonal corporation, often in disputes over land or water rights. In *Shane* (1953), the property rights of small farmers are threatened by large-scale cattle ranching. The allusion to the Johnson County War in 1892 was taken up explicitly by *Heaven's Gate* (1980), in which the moral ambiguities of the earlier movie are replaced by a Manichean struggle between big business and an ethnically mixed local populace. However, although the OK Corral has become an iconic location in western movies, the actual business of cattle ranching is usually subordinated to the moral and ethical dilemmas over male leadership that preoccupy the genre. Nevertheless, the cattle drives have been represented in the western genre as an integral part of the rise of the American nation. In *Red River* (1948), they feature as extended sequences of thrilling visual spectacle, central to the movie's celebration of an American empire based on the conquest of nature. Nevertheless, as Jane Tompkins observes, cattle in such movies are not seen 'for themselves', but only in relation to their utility for human beings, as 'factors in an economic scheme, as physical obstacles to be contended with in a heroic undertaking, or as the contested prize in an economic struggle'. This 'invisibility', she continues, 'is necessary if our society is to carry on some of its taken-for-granted activities: eating beef, wearing leather, using animal products, and continuing to support the huge and lucrative cattle industry—blood for money'.[1] An exception to the 'invisibility' of cattle in the western is the revisionist *Hud* (1963), which includes a sequence in which cattle infected with foot-and-mouth disease are herded into a pit and shot.

With his father facing financial ruin, Hud (Paul Newman) had cynically advised him to sell the cattle before the authorities find out they are diseased and have them condemned. The landscape of the Texas panhandle is photographed by James Wong Howe as flat, barren and inhospitable, a natural adversary against which the populist ideal of the virtuous family farmer is tested. However, although Hud's cynicism demolishes the myth of the heroic cowboy, the movie finally upholds the dignity and moral innocence of his father, a small farmer who refuses to give up ranching for the oil business because he says the latter occupation 'keeps a man doin' for himself'. The populist ideal of the self-reliant family farmer may be anachronistic in modern America, but its loss is nevertheless mourned in popular American film. This adherence to the myth of the morally innocent family farmer, however, tends to isolate Hollywood's farming narratives from a wider investigation of the destructive effects of modern agriculture, such as, in the case of cattle ranching, the damage done to grassland ecologies by the overgrazing of cattle.[2] Instead, the populist farming narrative tends to assume that the actions of the small family farmer are in harmony with the land, and, as the following section will also show, never a threat to its ecological health.

Survival and protest in the family farming narrative

Hollywood has perpetuated the populist myth of the yeoman farmer as a model of virtue, independence and democracy, living a rural life free from the contaminating influences of urban society. Accordingly, its farming narratives have drawn on a history of industrial monopolization, labour oppression and class conflict, Dust Bowls and crop failures, without placing these major transformations of the land within a discourse of ecological concern. Instead, the myth of innocence has been maintained either by showing farmers using benign, reassuringly old-fashioned technologies in an unproblematic way, or by omitting images of technology altogether. Moreover, in these human interest stories, nature is usually treated as a given, an unalterable fact against which the heroic farmers have to struggle in order to survive. The thematic emphasis of the films tends to be on gender relationships, and in particular the testing of patriarchal forms of masculinity, rugged, physically strong and individualistic, under conditions of social and economic pressure.[3]

The Grapes of Wrath (1940) established the formula for Hollywood's treatment of agrarian issues by celebrating the dignity and resilience of the Joad family, evicted from their homesteads in the Depression. The film's introductory title blames nature as the initial cause of the farmers' problems: 'In the central part of the United States of America lies a limited area called "The Dust Bowl", because of its lack of rain. Here drought and poverty combined to deprive many farmers of their land'. This account omits the contribution played by agricultural practices themselves in the creation of the Dust Bowls.[4] The next title, however, does include such cultural factors, stating that the farmer's family was 'driven from their fields by natural disasters and economic changes beyond anyone's control'.

The scene early in the film in which the man from the land and cattle company comes to evict Muley Graves confirms the priority of causative factors given in these titles. 'The fact of the matter is', he tells the farmer, 'after what them dusters done to the land, the tenant system don't work no more. It don't even break even, much less show a profit. Why, one man and a tractor can handle twelve or fourteen of these places. You just pay 'em a wage and take all of the crop'. 'Whose fault is it?', asks one of the farmers. In answer to this question, the film posits a complex system in which liability and blame are difficult to establish, much less act upon. The Shawnee Land and Cattle Company get their orders from the bank manager in Tulsa, who gets his orders from the East. 'Then who do we shoot?', asks Muley. The farmers are at the mercy of an impersonal system: the montage of caterpillar tractors evicting the farmers from their land reinforces the sense of mechanized forces acting impersonally and uncontrollably, and in marked contrast to the farmers' personal and emotional ties to the land.

The oppositions employed in *The Grapes of Wrath* are typical of the way in which agriculture has been represented in Hollywood cinema. The populist myth was later reinterpreted by *Easy Rider* (1969) as part of its celebration of an ideal 'counter-cultural' community. The biker couple's first encounter in the South West is with an Anglo-Mexican farming family which shows hospitality and a generosity of spirit towards them in marked contrast to the treatment they receive later in the small-town in the South, the intolerant values of which finally triumph in the film. The South-Western farmer is afforded respect by the outlaws for living an independent life, as Captain America (Peter Fonda) tells him solemnly: 'It's not every man that can live off the land, you know. You do your own thing in your own time. You should be proud'. The joking reference to the farmer's large family suggests that life close to nature is abundant, happy and fertile. At the end of the sequence, the bikers ride away to the music of The Byrds' 'I Wasn't Born To Follow', the lyrics of which reinforce the celebration of non-conformism found through closeness to the natural landscape of the American West.

The hippie commune at which the bikers stop next is also a place where people of goodwill are shown practicing small-scale, self-reliant agriculture in a life of rustic simplicity. The hippies tend a few sheep and goats, sow seeds by hand in arid soil, pray for the crops to grow, and give thanks for 'our place to make a stand'. Though the land seems barren, and Wyatt (Dennis Hopper) is sceptical about the commune's future, Captain America reassures him with the words 'They're gonna make it', as the movie again endows populist agrarianism with counter-cultural innocence. As they leave the commune, 'I Wasn't Born To Follow' is reprised on the soundtrack, to images of skinny-dipping hippies, playful and child-like. As Barbara Klinger observes, *Easy Rider* refuses to complicate its romantic construction of the rural South West with images of modernity comparable to the shots of gas stations, oil refineries and signs of African-American poverty that characterize its negative depiction of the reactionary South. The New Mexico commune, free of the corruptions associated with urban civilization, will continue to 'make it' on a frontier that is, apparently, still open. That the bikers' road trip into the restorative spaces of the rural South West is made possible by fetishized motor-bike travel and funded by a drug deal at LAX airport is a compromise with modernity which the movie also refuses to confront.[5]

A more nuanced and contradictory representation of farming occurs in *Days of Heaven* (1978), set in 1916, which departs from the idealized image of the itinerant farmer as a dignified innocent established by *The Grapes of Wrath*. The prairie landscape is photographed by Nestor Almendros as beautiful, rolling fields of wheat, often lit by golden light, in stark contrast to the noisy, smoky factory in which Bill (Richard Gere) works at the start of the film.[6] However, this tendency towards pastoral landscape imagery is subverted by the narrative's refusal to idealize farming. Although the *mise-en-scène* emphasizes nostalgic, picturesque shots of old agricultural machinery such as steam threshers and generators, the movie also shows the itinerant farm labourers as exploited victims of a class system, dependent for their living on the laws of supply and demand. 'They don't need you', Bill is told by a fellow worker, 'they can always get someone else'.

However, the narrative also tends to depoliticize this recognition of class conflict by displacing it into a morality tale of ambition, deception, love, jealousy, revenge and violence enacted in the triangular relationship between Bill, his sister Abby (Brooks Adams) and the rich landowner whom they try to defraud, known only as The Farmer (Sam Shepard). What is unusual about *Days of Heaven* is its attempt to locate this human drama within the wider context of the rural Texan landscape, through extensive use of close-ups of wildlife, including crickets, deer, rabbits, bison, partridges, grouse, cranes and herons. However, although given more emphasis than is usual in Hollywood movies, these foregrounded images of non-human nature still tend to function in a traditional way as signifiers of pastoral beauty. Moreover, because the animals are mostly photographed singly, rather than in relation to each other or to the human drama, the sense of ecological inter-relationship is undeveloped. Instead, the shots of non-human nature are mainly used in conventional terms as cutaways marking transitions in the human drama.

Non-human nature also serves its traditional purpose in *Days of Heaven* as objective correlative to The Farmer's mental and emotional state. The noisy invasion of the crickets is a sudden, unexplained natural phenomenon, which provokes The Farmer into setting his fields on fire to express his jealousy of Bill for his relationship with Abby. *Days of Heaven*, then, ultimately contains its representation of non-human nature within conventional Hollywood demands for a universalized, human interest melodrama.

The crisis in American farming in the 1980s provided topical material for two films released in 1984, *The River* and *Country*, both of which upheld the populist ideal of the resilient, dignified family farmer. As in *The Grapes of Wrath*, nature is established in both movies as a hostile force against which the farmer and his family must struggle. A cyclone at the start of *Country* convinces the Iowa farming family that it is in for another hard year ahead. The flood which begins and ends *The River* is equally given as an inevitable act of nature, caused simply by heavy rain or an act of God ('Oh Lord', comments Mae during the storm). The possible role played by human transformations of the Tennessee Valley, such as levée and dam building, is not mentioned in the film.

Both movies are built on the populist opposition of the honest, hard-working, stubbornly individualistic farmer against the corrupt interests of agribusiness corporations, politicians and banks. In *The River*, Joe Wade (Scott Glenn), boss of the Leutz Corporation, is introduced taking an aerial view of the flood damage from his helicopter, while the local Senator (Don Hood) is first seen in a shot that begins with his clean boots, as he gets into the helicopter and voices his resentment that the farmers are asking for more money. In stark contrast, Tom Garvey (Mel Gibson) and his family watch the helicopter from their flooded fields, covered in mud but closer to the honest earth than the representatives of big business and government.

The plot of *The River* centres on Wade's plan to buy out the farmers and flood the river valley for a dam. When the Senator asks him what will happen to the farmers in the valley, Wade replies: 'They're gonna get a fair price for their land. I'm not just talking about irrigation for myself. . . . I'm talking about enough hydro-electric power for the whole county. I'm talking about jobs. It could mean a lot to everybody. Including yourself'. As expressed here, Wade's motivations for buying Tom's farm combine his own economic self-interest with a recognition that the dam may also benefit the wider community. In keeping with the melodramatic mode, the movie adds to these rational motives a personal element: Wade is jealous of Tom for marrying Mae (Sissy Spacek). This personal motive confirms Wade's role as the film's villain, and enables audience allegiance to lie clearly with the farmer fighting to defend his traditional way of life against the bank's threat to foreclose on his loan.

As in *Days of Heaven*, the country/town opposition is heavily marked in the film. On several occasions, the movie cuts abruptly from the noise and smoke of the steel mill where Tom is forced to find scab labour, to the quiet, pastoral landscape of the farm, a landscape of beautiful sunsets prettified by Vilmos Zsigmond's soft-focus cinematography and John Williams' flute-based music.[7] This opposition climaxes in the sequence in which a stray whitetail buck accidentally runs into the factory, and is rounded up and set free by the workers, as it urinates in terror. Whereas industrial technology victimizes nature in this way, the agriculture practised by the Garvey family is, in contrast, benign and dignified, using reassuringly small-scale and old-fashioned technology. Mae hoes the ground and milks the cows by hand, while Tom's tractor is mocked by Harley as an 'old hunk of junk', but he cannot afford the new Kubota machine with air-conditioning and built-in cassette player. The images of the Garvey family crop-spraying are simply given, the ecological implications of such actions ignored. In this way, the movie establishes the Garvey family's convivial relationship with the land. 'This is our home place', Tom explains to Mae, echoing Muley Graves in *The Grapes of Wrath*. 'My people are buried here'. Although Mae argues that the family should move away from the river, she too has an emotional relationship with the farm, and weeps when the family's cow dies because they can no longer afford the vet.

When a black worker in the steel mill asks Tom, 'How come you grow all that food, and people still go hungry? There's something wrong, somewhere', Tom's reply is typical of the populist hero: 'I got no answers'. Tom's rebellion emerges from instinct and intuition, rather than from a political commitment or understanding of the economics of overproduction.[8] Again typical of the populist politics on which the movie draws, the individualist hero eventually learns the value of co-operation, as the Garvey family are joined by the rest of their community in an attempt to hold back the second flood with sandbags. Heroic masculinity, in crisis due to economic circumstances, is thereby restored: anxieties over the family's debt had made Tom sexually impotent, but the showdown with Wade at the end of the story gives him the opportunity to reassert his manly pride through armed resistance.

Nevertheless, the question of land development remains unresolved at the end of the film. Although the farmers have held out against the forces of agribusiness, their victory is only a temporary one, as Wade taunts them: 'Sooner or later, there's going to be too much rain, or too much drought, or too much corn. I can wait'. With his wealth and power, Wade can afford to concede this battle, and he ironically throws on the last sandbag to shore up the river bank. The forces of modernization seem as inevitable in the film as the rain and the flood.

The ambiguity and dramatic effectiveness of the sandbagging scene is, however, mitigated by the crude 'feelgood' tone of the film's coda, in which the Garvey children are shown playing in slow motion while the family harvests a golden field of corn. The movie thus cannot resist reasserting the agrarian myth of rural innocence: 'We're going to get a million dollars for all this corn', says one of the children before the end credits roll.

Country evokes a less romantic view of farming than *The River*, even though the movie's overall understanding of the farming crisis is similarly derived from populist agrarianism. As in *The River*, then, the impersonal forces of the credit system are again seen as responsible for the hardships of the American family farmer. Gleaner Ivy (Sam Shepard) is in debt to the Farmers Home Administration (FHA), and the bank forecloses on his debt.[9] In the past, the bank used to give loans on a personal basis, but it is now part of a chain, and decisions to loan money are made by the Loan Board in Des Moines. Gleaner comments that he can remember 'when this bank used to loan money on the man, not the numbers'. His reasons for not selling his land are also typical of the populist hero, and similar to those of Tom Garvey: 'This land's been in my family for over a hundred years'.

Encouraged by the FHA to borrow money in the boom period of the late 1970s, Gleaner accuses its representative, Tom McMullen (Matt Clark), of betrayal:

> Wasn't it you who was giving all these great speeches here a few years back, about, 'We're gonna feed the world', 'we're gonna expand, plant fence post to fence post', wasn't that you? Now here comes the government, put embargoes on foreign sales, and all these poor fools out here in the landscape with all their grain. No place to get a fair price for it.

Yet the movie dramatizes the farming crises in a nuanced way that avoids reducing it to Manichean simplicities. The possibility is raised, therefore, that the farmer himself may be partly responsible for his situation. 'You wanna know what's really got you?', McMullen replies. 'You're sitting on the most productive farmland in this country for a hundred years and you can't make a living on it. I didn't do that to you, you did that to yourself, my friend'. The suggestion of a personal failing on Gleaner's part is also voiced by his father-in-law Otis (Wilford Brimley). The notion of heroic innocence is thus complicated in this film, in comparison with the other farming narratives discussed elsewhere in this chapter.

As the debt situation worsens, one of the Ivys' farming neighbours commits suicide, and McMullen eventually resigns from the FHA. The narrative of *Country* is thus more sombre and less sentimental than *The River*, and does not evoke a pastoral landscape as a nostalgic escape. Indeed, landscape compositions in the film emphasize mundane farm buildings in a flat, monotonous landscape, usually filmed in wet, cold and cloudy weather. These are accompanied not by the pastoral music of *The River*, but by a simple, spare piano motif.

The narrative of *Country* is mainly concerned with tracking the emotional decline of the male farmer, under the pressures of debt and poverty. The family is forced to auction its farming equipment, and as Gleaner descends into drink and domestic violence against his son Carlisle (Levi L. Knebel) and his wife Jewell (Jessica Lange), the latter throws him out of the house. The family has been broken up by economic circumstances, and, as in *The River*, it is left to the farmer's wife to hold the family together. In both films, the wives are models of simple Christian piety, saying 'grace' before the family meals, and are strong and resourceful in the tradition of Ma Joad from *The Grapes of Wrath*, cooking, raising their children, and keeping the farm going while their husbands are absent. In *Country*, however, Jewell also becomes the focus of political resistance, organizing a community boycott of the FHA's auctions of farmers' property. At the end of the film, the Ivy family is reconciled, and it is reported on the radio that a federal judge has deferred further foreclosures until the farmers receive rights of due process. *Country* thus dramatizes the effects of economic, legal and political forces on the lives of individuals, in a way that avoids the sentimental evasions that compromise *The River*.[10]

The Milagro Beanfield War (1988) provides an interesting variation on the populist ideal of the yeoman farmer, being hailed at the time of its release as the 'first truly major, studio-financed film that deals directly with the Latino experience, featuring Latinos as leading, rather than secondary, characters'.[11] Adapting his 1972 novel for the screen, John Nichols told of his struggle to avoid pressure from the studio to stereotype his Hispanic characters and use big non-Hispanic stars.[12] The plot, updated to the Reagan era, concerns a conflict over land use and water rights in the fictional New Mexico village of Milagro. Profit-hungry land developer Ladd Devine (Richard Bradford) and State Governor (M. Emmet Walsh) plan to build a recreation area on land belonging to local Hispanic farmers.

When Joe Mondragon (Chick Vennera) accidentally kicks open an irrigation valve in his father's field, he decides to grow beans, using water controlled by the land developers. Stubborn and self-reliant, Mondragon refuses 'to go north like my dad to pick somebody else's beans for two fucking dollars an hour'. Yet traditional masculine heroics are less in crisis than in *The River* or *Country*: even Mondragon's enemies admit the man has 'huevos'.

The small farmer joins with the local community fighting to preserve its cultural heritage, and together they take on the system and win. As in *The River*, rural life is represented as idyllic, most notably in the sublime New Mexico sunsets which begin and end the movie. Milagro (Spanish for 'miracle') is the centre of humane values in the film: good neighbourliness, simple piety, humour, personal idiosyncrasy and collective support. These values are contrasted with the meretricious values of the citified property developers.

Whereas the farmers' technology is benign and small-scale (indeed, the villagers pick Mondragon's crop by hand), the forces of modernization are represented by the giant tractors tearing out trees and destroying the quiet tranquillity of Mondragon's beanfield, beautifully lit by the sun.

Ethnic differences play an important part in the narrative when resistance to the recreational development brings in white outsiders. Radical lawyer Charlie Bloom (John Heard) explains to the villagers how out-of-staters will take the new higher paid jobs and cause taxes to rise. Hurt by Ruby's accusation that he is a fellow traveller who will move out when the going gets tough, Bloom stays to play a vital leadership role in the village's fight.

However, the movie tends to depoliticize the villagers' resistance to the developers. The heroes of the movie, as reviewer Jill Kearney put it, 'don't seize history so much as bungle into it, and life proceeds by a process of sublime accidents'.[13] Indeed, director Robert Redford commented on the film that the 'idea of doing something without knowing why you're doing it, and finding out in the end, is more interesting to me than someone just setting out knowing. Most of the heroes in our culture are not aware that they're heroes'.[14]

The lack of overt politics in the film is reinforced by its reliance on the conventions of magic realism, which work against a materialist politics of land use. The movie includes two early scenes in which the Governor discusses the development of New Mexico with his advisers. Both of these scenes begin *in medias res*, and are therefore difficult for the viewer to follow, an alienation effect that reinforces the notion that a big business and political machine run by an unaccountable elite is affecting the lives of the innocent villagers.

As the movie develops, however, materialist explanations are downplayed, and mysticism proves to be an effective force against political and economic power. Saints and angels actually exist for the people of Milagro. Joe's dead father Coyote (Roberto Carricart) returns to old man Amarante (Carlos Riquelme) as a concertina-playing angel dressed like Pancho Villa. When the developers seize all copies of the local newspaper announcing a public meeting, Coyote magically summons a wind to blow them away into the sky. New York University anthropology student Herbie Platt (Daniel Stern), initially sceptical about local customs, eventually prays to St Jude, the patron saint of desperate causes, when Amarante falls ill.

The ending of the movie reinforces this wishful desire for a mystical resolution to economic and political crises. The audience's expectation of a violent resolution to the conflict increases when Coyote warns that the people are 'gonna need a big sacrifice here'. However, although the villagers buy guns to protect themselves, and undertake acts of sabotage against the leisure company, the expected armed confrontation does not happen. Instead, the armed stand-off between police and armed villagers ends peacefully. At the last moment, the Governor withdraws his plan to develop Mondragon's field, and orders his heavy Kyril Montana (Christopher Walken) via walkie-talkie not to arrest him. 'I'll explain it later', he tells him. The Governor then tells Devine: 'let's put it on the back burner'. In dramatic terms, the Governor's motivation seems unclear, and his surrender seems too easy. The narrative resolution is miraculous, indeed, as moral innocence is sentimentally rewarded and reaffirmed. The movie ends in an upbeat fashion, with a fiesta, at which even the local Sheriff (Ruben Blades) commits himself to the cause at the last moment, signing the petition even if it means he will lose his job. *The Milagro Beanfield War*, then, maintains the populist myth of the small, independent farmer who resists the destructive effects of modernity, and maintains a harmonious, nurturing relationship with his land.

Urbanization and power: *Chinatown*

Chinatown (1974) is based on the Los Angeles Aqueduct land fraud of 1905. In that year, engineer William Mulholland of the city's Water and Power Department announced plans for the building of an aqueduct to tap the water resources of the Owens Valley for the growing city of Los Angeles, over two hundred miles to the south. Covertly, Mulholland was conspiring with a syndicate of six Los Angeles millionaires, which was quietly buying land outside of the city in the San Fernando Valley. When the aqueduct was finished in 1913, the water did not go to the city, but to the San Fernando Valley, enabling the syndicate to sell off the newly improved land at a great profit. Meanwhile, farmers in the Owens Valley were left without water.[15]

Director Roman Polanski described *Chinatown* as 'showing how the history and boundaries of L.A. had been fashioned by human greed'.[16] The movie does include events found in the historical record, notably the secret dumping of water into the city's sewer system to create an artificial water shortage, and thereby force the citizens of Los Angeles to vote for the public bond issue to finance the aqueduct. However, Robert Towne's script transposes the land grab to the 1930s, thereby reworking it as a revisionist detective and murder mystery, in which small-time matrimonial detective Jake Gittes (Jack Nicholson) attempts to discover who is behind the conspiracy involving the city's Water and Power Department. This transposition not only turns the historical source into a 'pitchable genre yarn', as Michael Eaton puts it, but also has the effect of dehistoricizing the narrative, so that it becomes an inquiry into human corruption rendered timeless and universal.[17] Indeed, for Jacoba Atlas, writing in the *Los Angeles Free Press* after the release of the film in the summer of 1974, the story of the land grab, with its basis in what she called 'the basic conspiratorial trauma of our past', was analogous to the political situation of her own time. A 'rundown of the tactics used to secure the water for the San Fernando Valley and to annex that portion to the City', she wrote, 'reads like a parallel study to the tactics used in the Watergate scandal of today'.[18] In *Chinatown*, then, the 'Rape of the Owens Valley' is posited as a universal archetype for the corruption on which modern America is built and maintained, corruption which, as the following analysis of the film will show, is ultimately an expression for Towne and Polanski of original sin, timeless and inevitable.

The mediation of American history by the detective genre crucially shapes the environmental politics of *Chinatown*. The information disclosed by the detection scenario is itself fairly predictable, in that the viewer learns early on that the Water and Power Department is involved in fraud and corruption, and suspects the involvement of Noah Cross (John Huston). The surprise withheld to the end of the film is the secret that Noah's daughter Evelyn (Faye Dunaway) is holding: that she is the victim of her father's incest, and that Noah is therefore a sexual, as well as a social, criminal. This doubling of Noah's rape of the land with the incestuous rape of his daughter turns the land grab into an exemplum of a wider, more pervasive evil. Robert Towne commented: 'Maybe it's because . . . America's a puritanical country I felt that the way to drive home the outrage about water and power was to . . . cap it with incest'.[19] However this doubling tends to depoliticise the film, in that the environmental crime becomes no longer specific to a particular time and place, but the consequence of a corrupt human nature that is universal and innate. 'The taint of corruption is exposed not as the measure of a specifically dysfunctional society,' writes Michael Eaton, 'but, in fact, forms the foundation of all human interaction'.[20] The specific history of the Los Angeles land fraud has, moreover, been turned into a more sensational melodrama of murder and sexual violence.

The ending of the film subverts the normative expectations of the detective genre, in that the final revelation of truth leads not to a valorization of rationality and justice, but to their defeat. The crimes of the city fathers go unpunished, and, however well intentioned, the protagonist's actions have tragic, unforeseen and uncontrollable consequences, as Gittes' attempts to save Evelyn lead to her death. The overall sense, therefore, is that human nature is essentially corrupt, so that political resistance is futile. Unlike Philip Marlowe or Sam Spade, Gittes is unable to gain control of events, and is defeated by the enormity and impersonality of state and business power. The ending of *Chinatown* thus refuses the affirmative closure familiar to the melodramatic mode, but as a result ends on a note of fatalistic resignation. As the next section will show, Hollywood critiques of urban society within the science fiction genre, in contrast, tend to mitigate their apocalyptic premises with melodramatic gestures of hope.[21]

The total city and the end of nature: *Soylent Green*

Hollywood cinema, as Colin McArthur shows, has had an ambiguous relationship with the city, both celebrating and criticizing it.[22] Few movies have conceived of the city in ecological terms, however, though science fiction movies have tended to be exceptions.

A recurrent myth in recent science fiction is that non-human nature has ended, destroyed by a closed technological environment in which urbanization and commodification are total. In the dystopian New York of *Soylent Green* (1973), set fifty years in the future, the city has suffered ecological catastrophe, and nature is dead. The air is polluted with a yellow smog, and there is a permanent heatwave due to a greenhouse effect. The population of the city has reached forty-one million. Most are unemployed, seen squatting passively in doorways, hallways, subway cars and parking lots. Power shortages mean there is no industry, while labour relations are poor: detective hero Thorn (Charlton Heston) will lose his job if he is sick for two days. In these conditions of scarcity, the cost of real food is high, only affordable by the rich elite. Instead, the Soylent Company, which controls the food supply for half the world, provides the masses with tasteless, synthetic food.

The narrative takes the form of a detective quest, the revelation at the end of which is that Soylent Green, a cracker which the company claims is made from plankton, is actually made from human corpses. Monopoly capitalism has literally consumed the human race itself. The film, then, is a response to anxieties about overpopulation which came to public attention with the publication of Paul Ehrlich's *The Population Bomb* (1968), a Malthusian prediction of famine and environmental collapse in the United States by the 1980s. In *Soylent Green*, anxieties about overpopulation are marked especially by fears of standardization and homogenization, suggesting that middle-class anxieties over *déclassement* are at the centre of the narrative, rather than the racial anxieties that so often mark arguments about overpopulation.

Against the ruins of ecological disaster, *Soylent Green* mourns the lost authenticity provided by three obsolete resources: religion, the written word and non-human nature. Sol (Edward G. Robinson) is the dignified upholder of these lost values. He blames the 'scientific-magicians' who have 'poisoned the water, polluted the soil, decimated plant and animal life' for the death of nature. When he discovers the truth about Soylent Green, he checks into a voluntary euthanasia clinic, where he dies watching a nature movie edited to Beethoven's Pastoral Symphony. The detective Thorn, cynically used to the present conditions, has his complacency challenged when he sees these images of lost nature: deer, daffodils, rivers, undersea fish. 'How could I know?' he weeps. 'How could I ever imagine?' Ultimately, however, the movie depoliticizes its environmental message by locating blame in a universally corrupt human nature. 'When you were young, people were better', Thorn suggests to Sol at one point. 'Nuts', the latter replies. 'People were always rotten, but the world *was* beautiful'.

Soylent Green locates possible hope for future change in the values of literacy and scholarship which Sol helps to preserve in a secret library. In the final scene of the film, Chief of Police Hatcher (Brock Peters) implores Thorn to 'tell everybody . . . Soylent Green is people'. To a reprise of the Pastoral Symphony, Thorn's bloody, upraised finger dissolves into an image of tulips in a field, as the movie ends with a recapitulation of the nature images which Sol watched as he committed suicide. Given that the dystopian city has been established as a totally closed environment, however, such images appear merely nostalgic, and the affirmation of resistance centred on Thorn's upraised finger an empty rhetorical gesture. Although *Soylent Green* builds its narrative on environmentalist premises, therefore, its formulation of ecological crisis as already total, and of corporate and state power as monolithic, leaves little space for the formulation of a convincing politics of resistance. Genre conventions again play a vital part in shaping the politics of a Hollywood environmentalist movie.

ELEVEN

The Ecology of Automobile Culture

The automobile is not simply a utilitarian machine, but an irrational vehicle for what Eric Mottram calls 'the conversion of transport into libidinal impulse'.[1] Hollywood cinema has long perpetuated the myth of the automobile as a symbol of the triumph of American capitalist values of possessive materialism and individual freedom defined as mobility. The car chase, staple of the thriller genre, fetishizes thrills of speed and risk at the edge of control.[2] *American Graffiti* (1973) celebrated the 1950s as a utopia combining the erotic pleasures of youth, neon light, rock and roll music and the fetishized bodies of automobiles. But the movie's location of such fantasies in the past is as signifi-cant as its elision of their ecological consequences. In contrast, the opening scenes of *Falling Down* (1992) replayed images, increasingly familiar from the late 1960s, of the 'freeways' of Los Angeles as places of constraint, in which traffic jams, heat haze, claustrophobia and failing tempers denote the social and technological breakdown of the automobile-dependent American city.

Like non-human nature, automobiles have mostly been taken for granted in Hollywood films. This chapter, however, explores two areas in which environmental issues surrounding the automobile have been relatively visible. The first is in what David Nye refers to as 'energy narratives' dealing with the quest for fuel.[3] The second is in movies that involve the search for alternative technologies to improve or replace the automobile itself. In these movies, American history has been represented as a conspiracy of automobile and energy cartels, developers and industrialists, plotting to suppress alternative technologies that might improve or replace fossil-fuel consumption and the polluting automobile.

The quest for oil

The Formula (1980) is what David Nye calls a narrative of 'artificial scarcity', in which 'a person or institution creates scarcity for personal gain'.[4] An international oil cartel, led by Adam Steiffel (Marlon Brando), is suppressing a secret formula developed by Nazi scientists for the clean extraction of gasoline from coal. 'You're not in the oil business, you're in the oil scarcity business', detective Barney Caine (George C. Scott) tells Steiffel. In the end, the international conspiracy continues its greedy search for profit maximization, and Caine resigns himself to being powerless to change the system. 'I'm not an adversary any more', he tells his police colleague, 'I'm just another customer.' The movie ends on a freeze-frame of traffic building up on an urban interchange. The political situation at the end of this convoluted narrative is equally static and fixed.

A similar sense of political impasse is suggested in *The Two Jakes* (1990), the second film in Robert Towne's projected trilogy about Los Angeles, in which the screenwriter extended his exploration of land use in the city from the corrupt speculation in water he dealt with in *Chinatown* (discussed in the previous chapter), to real estate, oil exploration and highway construction. Set in 1948, the capitalist history of Los Angeles repeats itself as another labyrinthine conspiracy involving public and private corruption, adultery, murder, intimidation and blackmail, or as Jake Gittes (Jack Nicholson) muses, 'old secrets, family, property, and a guy doing his partner dirt'.

Power is now in the hands of Texan oil company boss Earl Raleigh, who is secretly drilling for oil in an earthquake zone. At one point, Raleigh confides his love for the 'smell of sulphur': the fossil-fuel producer as Satan. The plot involves the plan by real-estate dealers Berman and Bodine to develop the same valley for a suburban housing estate dependent on the automobile, even though the valley is at risk from gas explosions from the oil drilling. Raleigh explains to Gittes how the suburbanization of Los Angeles will depend on the oil he owns:

> Without my oil, you've got no automobiles. Without automobiles, you've got no road construction, no sidewalks, no city lights, no gas stations, no automotive service, and no Berman subdivision stuck out in the toolies because nobody can get there. Mr. Berman's out of business before he even gets in business. The name of the game is oil, John.

When Raleigh assures Gittes that he is drilling for oil out at sea, and not in the valley, the movie exposes him as a liar almost immediately. For as in *Chinatown*, *The Two Jakes* is less interested in generating narrative mystery around the issue of public corruption, than in withholding information on the personal motivations of its central protagonists. In this case, Gittes discovers that Berman shot Bodine not to gain the land rights to the valley, as he originally suspected, but to protect his wife Kathryn from blackmail. The important revelations (akin to the revelation of incest withheld to the end of

Chinatown) are that Kathryn is the child of Noah Cross's rape of his own daughter Evelyn (in the first movie), and that Berman is himself dying of cancer. The two Jakes (Gittes and Berman) are thus both attempting, for their own reasons, to protect the same woman, Kathryn, from scandal and distress.

The return of the incest theme means that the power conspiracy involving the modernization of Los Angeles is again represented as psycho-sexual in origin, in a way that tends to posit corruption as an essential and unchangeable part of human nature. The sexual basis of Raleigh's desire to exploit the land for its oil is revealed in his conversation with Gittes, when he compares drilling for oil to the horse breeding practice of 'whipstocking'. 'Well it's sorta like helping a stallion mount a mare', he explains. 'Whipstocking is something you do to coax the drilling bit in the right direction. After you've gone to all that time and trouble, you wouldn't want your big fella to miss what he was aiming at, now would you?' Raleigh's comparison of the exploitation of the land with an act of sexual penetration recalls the doubling of Noah Cross's land grab and his rape of his daughter Evelyn in *Chinatown*. In both films, the male sexual desire to dominate a feminized nature is seen as the main motive for the villain's destructive use of the natural environment.

In *The Two Jakes*, Gittes finally identifies Kathryn by her necklace of wild poppies that changes colour. Kathryn, like nature itself, is an unknowable mystery that needs to be protected from the threat of male violence. However, Kathryn is already lost, cursed by a past that, as she recognizes at the end, 'never goes away'. The weight of original sin, then, weighs heavily in *The Two Jakes*, as in *Chinatown*. Compared to his failure at the end of the first movie, however, Gittes this time is able to act effectively to protect the female, when he lies at the preliminary hearing into Bodine's murder and tampers with an incriminating tape recording in order to keep Kathryn's name out of the trial.

Nevertheless, Gittes remains powerless in the face of the wider corruption that his investigations reveal, unable to undo his past mistakes or do anything to change the future, and himself complicit in the very corruption he reveals, as his sexual desire to dominate Bodine's widow Lillian suggests. No one can stop Raleigh and Bodine, he tells city engineer Tyrone Ottley, before concluding: 'Guys like Raleigh don't get arrested, they get streets named after them. This type of situation, it's best just to leave it alone. Raleigh is stealing from Berman. It's a big thief stealing from a little thief. Who are we to quarrel?' The sense of melancholy resignation is therefore pronounced, as it was at the end of *Chinatown*. That *The Two Jakes* similarly naturalizes, and therefore depoliticizes, the predatory nature of human beings, is made clear in the reference to Darwinian nature in the lecture given by Ottley at the La Brea Tar Pits, in which he speaks of 'animal after animal literally dying to eat another dying animal'. This model of nature as a scene of Darwinistic competition and violence confirms the view of human society in both *Chinatown* and *The Two Jakes*. Both movies refuse to take recourse in the populist values that give many environmentalist movies their optimistic resolutions. In doing so, however, they are unable to formulate strategies of resistance to the processes of modernization and development they construct as inevitable and unchangeable.

In comparison, *The Pelican Brief* (1993), based on John Grisham's novel, is a conspiracy thriller that ends with a reassuring affirmation of the efficacy of the American legal system to win justice for ordinary people. Financier Victor Monteith, wanting to exploit a wetlands site in Louisiana for oil, despite the protests from environmental pressure groups, has the two Supreme Court judges who overrule his claim assassinated. In the end, however, law student Darby Shaw (Julia Roberts) and investigative newspaper reporter Gray Grantham (Denzil Washington) successfully uncover the truth, with the help of audio and video tape evidence of the conspiracy. With Monteith indicted, Darby Shaw is flown to a secret island location under the protection of the FBI, and the movie ends with Julia Roberts' smile. Although the President comes out of the story in a negative light, after his attempt to cover up his associations with Monteith, in *The Pelican Brief* ordinary people triumph over the corruptions of the oil business, protected by the benign forces of official law and justice.

In conformity to the requirements of the action-adventure genre, *On Deadly Ground* (1994) produces an equally melodramatic representation of the issues of oil extraction. The villainous oil company executive Jennings, played with nostril-flaring excess by Michael Caine, is motivated solely by greed, and employs ruthlessly violent methods for the sake of private profit, with no regard for health and safety considerations. Aegis Oil knowingly uses substandard equipment to hurry through its drilling programme before the oil rights revert to the local Native Alaskans. The whistleblower to the Environmental Protection Agency is killed by Jennings' sadistic henchman. Forrest Taft (Steven Seagal) discovers the full extent of the company's moral corruption: the faulty 'preventers' they have negligently used in the construction of the rig mean that when it goes on-line, it will explode. The company plans to dump its waste products, and broker the space to another company, before the explosion.

However melodramatic, such negative representations of big oil corporations identify a source of widespread popular suspicion and resentment. However, the movie's politics of melodramatic gesture were contested by interested parties on all sides of the conflict over land use in Alaska. A staff biologist for Greenpeace found the movie not only too violent for the pacifist sensibilities of most environmentalists, but also 'so simplified and hokey. In real life, the industry is much more subtle and slippery.' Moreover, she pointed out that oil companies prefer to exercise the permissions granted by law, rather than simply act illegally, a subtlety that the film misses. In contrast, a spokesman for British Petroleum defended the reputation of the oil industry: 'Any conception of the oil industry as having no regard for the Natives, their beliefs, or the environment, and being only profit-motivated . . . is clearly a misconception', he told *The Los Angeles Times*.[5]

Both of these interpretations of the movie judge it on realist principles, and conclude that it does not explore the ethical and political issues it raises with a satisfactory degree of complexity. Indeed, judged from this realist perspective, *On Deadly Ground* travesties the complexity of the debates over land use in Alaska. Important issues, such as the role of state and federal government, are simply ignored or given token mention. Moreover, the political debate is presented in a way that is both oversimplified and inaccurately polarized. As Roderick Nash argues, environmental issues in Alaska are not reducible to an either/or choice between wilderness preservation and industrial use. Not all Native Alaskans are wholly against the oil industry, as the film tends patronizingly to assume. Native opposition to oil production in Alaska remains focused on its harmful impact on subsistence hunting and fishing. However, under the Alaska Native Claims Settlement Act of December 1971, Native Alaskans gained an economic interest in oil production, so that opinion on the issue of oil leasing in the Arctic National Wildlife Refuge is divided. Some Natives see the oil industry as their benefactor, and are keen to sell the exploration and extraction rights for oil on Native-owned land. The Inupiat, observes Shepard Krech III, vary from those opposed to further drilling in the Refuge, to others who proclaimed their support for the oil industry in slogans such as 'thumbs up for development' and 'oil is the future'.[6] *On Deadly Ground* ignores such complexities, which would seriously compromise its sentimental adherence to the figure of the ecological Native American discussed in Chapter Four.[7]

Ultimately, the movie opts instead for the technological fix as a solution to the predations of the oil industry. At one point, pursued by Jennings' posse, Forrest Taft abandons his helicopter for a horse. If this action suggests an ambivalence towards big technology, then the narrative is resolved through the hero's recourse to what reviewer Tom Tunney described as 'all the paraphernalia of jungle guerrilla combat: deadly booby traps, claymore mines, an M-16 rifle and plastic explosives'.[8] As mentioned in Chapter Four, this use of heavy weaponry confirms the hero's rejection of the pacifist spirituality of the Native Alaskans. Moreover, by fetishizing the techno-fix, the movie conforms to the familiar expectations and addictions of the action-adventure genre, providing a spectacle of both technology and the male body in violent display. Of course, this equation of technological mastery over the environment with masculine prowess is viewed by many radical environmentalists, especially ecofeminists, as a central cause of ecological destructiveness.[9] Ironically, then, from this perspective, Forrest Taft preserves the Alaskan wilderness by deploying the very strategies of masculinist, aggressive mastery through technology that have contributed to its degradation in the first place.

The long polemical speech which Forrest delivers to the tribal council at the end of the movie endorses the development of alternative technologies as a solution to the problem of fossil-fuel pollution, and thereby displays the movie's alignment with mainstream environmentalism. The speech advocates research into new 'alternative engines' for automobiles in order to reduce dependence on oil extraction. Forrest argues that the vested interests of the oil cartels are combining with 'corrupt government regulations' to suppress the fact that 'the concept of the internal combustion engine has been obsolete for over fifty years'. The movie's liberal reformist solution thus brings together faith in technological progress with an appeal to democracy and the rule of law:

> I think we need a responsible body of people that can actually represent us rather than big business. This body of people must not allow the introduction of anything into our environment that is not absolutely biodegradable or able to be chemically neutralised upon production. And finally as long as there is profit to be made from the polluting of our earth, companies and individuals will continue to do what they want. We have to force these companies to operate safely and responsibly and with all our best interests in mind, so that when they don't, we can take back our resources and our hearts and our minds and do what's right.

Politically, the speech deploys its populist rhetoric (appealing to sentiment, and polarizing the situation as 'us versus them') to argue for a more regulated form of capitalism, in particular the tightening of legal regulations on toxic dumping. It evokes the seductive authority of scientific jargon (products must be 'absolutely biodegradable' and able to be 'chemically neutralised upon production') to guarantee a technocentric, rather than ecocentric, form of environmentalism, reflecting, as David Pepper puts it, belief 'in the existing structure of political power, but [with] a demand for more responsiveness and accountability in political, regulatory, planning and educational institutions'.[10]

This advocacy of 'green' technology by *On Deadly Ground* has its limitations, however. In particular, 'green' engines may make automobiles cleaner, but will not solve the problem of gridlock, or the pressure that road building places on the conservation of rural or wilderness areas. From a more radical environmentalist perspective, then, over-reliance on the techno-fix contributes to the greenwashing of environmental problems, by offering evasive, simplistic and apolitical answers to a complex set of issues. On the other hand, as Gross and Levitt rightly comment, technological developments should not be idly dismissed as 'techno-fixes', for, as they put it, 'a 1 percent improvement in the efficiency of photo-voltaic cells, say, is, in environmental terms, worth substantially more than all the utopian eco-babble ever published'.[11]

Although the mainstream environmentalist values embraced by *On Deadly Ground* do not fundamentally challenge capitalism, the movie does question whether economic exploitation and growth are necessary and inevitable, and whether profit should be made, as Forrest Taft puts it, 'from the polluting of our earth'. The movie therefore nostalgically constructs Alaska as the last frontier, a wilderness to be preserved apart from its economic use value and the invasiveness of modern technology. The final image of the film is of the white hero and his Native American female companion sitting in a kayak on a placid lake in the mountains, in a pristine natural landscape in which modern technology is absent. The dwarfing of the protagonists by the sublime vastness of the lake and mountains gives a sense of their vulnerability: nature continues to be under threat, and the future of the Alaskan wilderness not neatly resolved.

Free Willy 2 (1995) also promotes a mainstream environmentalist agenda regarding the oil industry, while representing its executives as cynical and profit-minded. Having irresponsibly caused an oil spill, they hide deceitfully behind their reassuring public relations announcements, while secretly conspiring to display the orcas caught up in the spillage for profit at a marine rescue centre. Wilcox thinks it will be a success, 'As long as it looks like we have the whale's best interests at heart.'

Despite this negative view of the company executives, however, the criticism of the oil industry made by *Free Willy 2* is moderate. There are suggestions that commercial imperatives are to blame for the tanker running aground: running late, the captain orders it to travel too fast, without a tug escort, in dangerous conditions. Yet the fundamental environmental problem is given as technological, rather than economic or political. As television presenter Joan Lunden explains on *Good Morning America*, the single hull tanker 'was built well before 1990, the year when double hulls became mandatory for all new vessels, and that many feel could go a long way to prevent disasters like this one'. The voice of a protester confirms this reassuring argument: 'We need double-hull tankers'.[12]

This positing of a technological solution to environmental hazards aligns the movie with mainstream environmentalism, in that it advocates that the transportation of crude oil be made safer, while avoiding more radical questions connecting energy consumption with economic growth and political power. As with *On Deadly Ground*, the techno-fix is also appropriate for the action-adventure genre to which *Free Willy 2* belongs. Unusually for the genre, however, the narrative resolution relies on a combination of technology and mysticism. When one of the orcas is poisoned by the oil spill, the Haida Indian Randolph, representing a more benign attitude to nature than that of Benbrook Oil, provides traditional natural medicine which works on the sick animals better than the scientific medicine of Dr Haley. Randolph then gives Jesse a wooden charm that will guarantee that the orca, as he puts it, is 'close to your spirit'. Jesse is thereby initiated into a mystical relationship with the orca, a bond that saves his life when he summons Willy to save him from drowning. More predictably for the action-adventure genre, a helicopter then winches Jesse and his friends to safety. In *Free Willy 2*, then, the technological fix is reconciled with New Age mysticism, and the moral innocence of the orca, the Native American and the boy is opposed to the corrupt machinations of the oil company.

'It's the idea that counts, and the dream': challenging the automobile cartel

As the title of Francis Ford Coppola's biopic of American automobile designer Preston Tucker suggests, *Tucker: The Man and His Dream* (1988) celebrates Tucker (Jeff Bridges) as a maverick inventor taking on the vested interests of the Big Three Detroit automobile manufacturers. Tucker stands for a reformist version of the true American Way, exuding confidence, risk-taking, ambition, and dedication to both hard work and his family. But his populist dream to build his 'revolutionary' new car is eventually thwarted by the combined forces of corporate monopoly, Washington politics and the press. The movie identifies Tucker as a martyr: when he talks about 'Having ringside seats at your own crucifixion', the image cuts to a shot of the 'T' in the factory's 'Tucker' sign being raised like a cross.

When he fails to build his new cars on time, the Securities and Exchange Commission charge Tucker with fraud. But his lawyer Abe (Martin Landau) tells him the real reason for bringing him to trial: 'you made the car too good'. However, at the court hearing which climaxes the film, Tucker convinces the jury that he took public money in good faith, and made an honest attempt to make the cars. The maverick inventor walks free.

The big automobile manufacturers are shown to be too conservative to experiment with technological innovation. Robert Bennington, former president at Plymouth, tells the court hearing: 'A well run corporation doesn't waste money to research innovations, unless, of course, keeping up with the competition demands it.' Yet the movie rejects Bennington's caution, and is nostalgic for the pioneering days of risk-taking venture capital. As Tucker himself puts it: 'It's the idea that counts, and the dream.'

However, in environmentalist terms, Tucker's 'revolutionary' 'car of tomorrow' was really only a moderate reform of the existing paradigm, improving performance and safety, rather than adding environmentally friendly controls. The innovations combined fuel economy (20 miles per gallon) with speed (130 miles per hour top speed, streamlining, improved aerodynamics) and pioneering safety aspects (rear engine, a windshield that pops out, fenders and headlights that turn with the wheel, shatterproof glass, seatbelts, and a sponge rubber crash panel).[13] Such innovations in safety and economy were important for their day, but, from a contemporary perspective, did not address more fundamental, long-term environmental issues such as gridlock, toxic emissions and dependency on fossil-fuel consumption. Indeed, as the Afterword to the movie shows, all of Tucker's innovations were eventually adopted by the Detroit companies themselves. The Tucker car was, therefore, a technological fix that helped to maintain, rather than challenge, the existing paradigm of automobile culture. Moreover, Coppola's movie does the same, by celebrating the glamour of car culture, in all of its aspects: advertising, giganticism, speed and commodity fetishism. Tucker brags enthusiastically that his car will be made in the biggest factory in the world. The inventor's zest for life and rule-breaking non-conformism are also best expressed in the car chase with the police, cut to the syncopated accompaniment of 'Where's That Tiger', a sequence which celebrates the pleasures and exuberance of automotive speed. The 1950's automobiles are seductively photographed by Vittorio Storaro as clean, shiny machines, with a loving attention to visual detail.

A more critical view of automobile culture was taken by the mildly satiric family movie *Who Framed Roger Rabbit* (1988), released in the same year as *Tucker*. Los Angeles in 1947 is introduced as a double utopia: it is 'Toontown', the place where cartoon fantasies come to life, and also a city where public transportation is cheap and reliable. 'Who needs a car in L.A.?', asks Eddie Valiant (Bob Hoskins) with dramatic irony, 'We got the best public transportation system in the world.'

But the utopia is under threat from real-estate dealer Judge Doom (Christopher Lloyd), who mistakes his own murderous greed for a progressive vision of the city's future. He buys the Red Car Trolley company with the aim of dismantling it, in order to force an automobile economy onto Los Angeles. In this way, the movie alludes to the conspiracy theory surrounding the Yellow Truck and Coach company, 50 per cent owned by General Motors, which in the 1930s began to replace existing tramlines with bus services, and thereby helped bring about the decline in mass transportation in the city.[14]

The plot of the movie involves Judge Doom's attempt to develop Toontown for a freeway. As he tells Eddie: 'Several months ago I had the good providence to stumble upon a plan of the city councils—a construction plan of epic proportions. They are calling it a "freeway". . . . Eight lanes of shimmering cement running from here to Pasadena. Smooth, safe, fast. Traffic jams will be a thing of the past.' When Eddie questions the need for freeways in a city with good public transportation, Judge Doom replies: 'You lack vision. . . . I see a place where people get on and off the freeway, on and off, off and on, all day, all night. Soon, where Toontown once stood, will be a string of gas stations, inexpensive motels, restaurants that serve rapidly prepared food, tyre salons, automobile dealerships, and wonderful, wonderful billboards reaching as far as the eye can see. My God, it'll be beautiful.' In the end, however, Roger Rabbit's 'sense of moral outrage' triumphs over cynical big business, and Toontown is granted in perpetuity to the lovable 'toons. The movie thereby offers a cartoon solution to the real problems to which it alludes.

An uneasy but interesting combination of political satire and romantic comedy, *The American President* (1995) also provides a Hollywood utopian fantasy, in its assertion that institutional environmentalism, in alliance with a strong, liberal presidency, can produce progress on toxic emissions, despite the compromises and dirty dealings of Washington politics.

Democratic President Andrew Shepherd (Michael Douglas) is preparing two controversial pieces of legislation, a Crime Bill and an Energy Bill. The latter, Resolution 455, includes the target of a 20 per cent reduction in fossil-fuel emissions to tackle global warming. The President's advisers suggest he lower the target to 10 per cent to secure support for his Crime Bill from three senators from California and Michigan with automobile interests, who would otherwise vote against it. However, the President promises Sydney Ellen Wade (Annette Bening), lobbyist for the Global Defense Council, that he will support the more radical 20 per cent reduction in fossil-fuel emissions if she secures the necessary votes.

When the widowed President begins a sexual relationship with Sydney, the narrative centres on their attempts to keep their personal and public lives separate, while Republican Senator Bob Rumson (Richard Dreyfuss) exploits what he sees as a conflict of interest. However, when Sydney inadvertently lets slip at the White House Christmas party details of her negotiations with the three senators with automobile interests, the President's Chief of Staff (Martin Sheen) also raises the issue of conflict of interest. 'Did the GDC's political director just tell the President and the White House Chief of Staff that there are three votes on the Crime Bill that can be bought by sticking the fossil fuel package in a draw?', he asks. 'No', replies the President, 'the GDC's political director didn't tell us anything. Sydney Wade told her boyfriend and her boyfriend's best friend that she had a lousy day.'

Nevertheless, the President acts on the information divulged by Sydney to dilute the Energy Bill, in order to get his Crime Bill passed. When she finds out, Sydney feels betrayed, and breaks off their relationship. The

President then calls an unscheduled press conference, at which he announces his decision both to cancel the Crime Bill, in order to redraft a stronger version that includes legislation on hand gun control, and to submit the more radical Energy Bill to Congress. 'America isn't easy', he tells the impromptu press conference, 'America is advanced citizenship', before ending his speech with a reassertion of his authority: 'I *am* the President'. Happily reunited, both Sydney and the President maintain that they did not act out of personal reasons. 'I didn't send 455 to the floor to get you back', he tells her. 'And I didn't come back because you sent 455 to the floor', she replies.

Nevertheless, the movie's resolution of the conflict between private life and public duty is an uneasy one, as reviewer John Wrathall noted: 'Shepherd's last-minute decision to go all out for the Energy Bill . . . suggests that perhaps Rumson was right all along, and that the President *has* allowed his policies to be unduly influenced by his personal life'.[15] The film's generic mixing of romantic comedy and political satire thus means that the happy resolution of the love story produces a confused and incoherent resolution to the political subplot. Moreover, in a historical context of reluctance on the part of successive Presidents to back a strong policy on toxic emissions, the idea of an American President agreeing to a 20 per cent reduction in fossil-fuel emissions in ten years, is, of course, pure Hollywood utopianism.[16]

TWELVE
The Risks of Nuclear Power

From its inception in the 1940s, the American nuclear industry has promoted itself with imagery evoking a machine in the garden producing a clean, white power.[1] However, Hollywood cinema has tended to reject this utopian narrative of a liberatory science and technology, representing the effects of nuclear technology, both domestic and military, in more sensationalist and emotive ways, in countless variations on the Frankenstein narrative, in which the Promethean attempt by a mad or evil scientist to master nature is an act of hubris that invites punishment. Narratives in which nuclear technology is inherently unstable, or is abused when it falls into the wrong hands, have become commonplace. After a brief survey of Hollywood movies concerned with the military uses of nuclear power, this chapter will analyse in detail the representation of the domestic nuclear industry in the United States in two popular films, *The China Syndrome* (1978) and *Silkwood* (1983).

In the 1950s, science fiction movies visualized the unfamiliar and potentially catastrophic effects of nuclear testing and atomic radiation in dystopian images of invasion by biological mutations, symbolizing the revenge of nature against human transgression. In *Them!* (1954), Los Angeles was terrorized by giant ants produced by atomic tests in the desert; the following year, in *It Came From Beneath the Sea* (1955), a giant octopus, also a product of nuclear mutation, attacked San Francisco. *On the Beach* (1959), in contrast, exemplified a new trend in more realist speculations on future dystopian scenarios, a cycle that continued into the Kennedy era with *Seven Days in May* (1964) and *Fail-Safe* (1964).[2]

With the renewal of the Cold War in the early 1980s, popular anxieties about nuclear war led Hollywood to revive the nuclear war movie. The low-budget *Testament* (1983) elided political issues to explore the emotional effects of nuclear attack on a close-knit community in California. In contrast to such psychological understatement, thrillers of the period gratefully took the plot device of nuclear material falling into enemy hands as their 'Macguffin'. *Thunder Run* (1985), for example, was a trucker movie with an aggressively Reaganite agenda. Korean war veteran Charlie Morrison comes out of retirement to drive a dangerous consignment of plutonium for the government. The enemy is a terrorist group whose leader is called Carlos. Morrison deals with both the political threat of terrorism and the personal problems within his own family, as son follows father in a masculine rite of passage through violence. Plutonium, the movie assures the viewer, is safe if in the right hands, understood as those of the hegemonic, white American male.

In the post-Cold War era, anxieties that stray nuclear weapons may get into the hands of those defined by the United States as terrorists continued to provide plot lines for thrillers such as *True Lies* (1994), in which the enemy is a caricatured 'Arab' terrorist group Crimson Jihad, and *Crimson Tide* (1995), an old script from the Cold War era reworked with an extreme Russian nationalist replacing the Soviets as the enemy. Nuclear weaponry thus continues to provide action-adventure and war movies with apocalyptic thrills of danger, survival and male heroism.

Anti-nuclear discourses, notes Spencer Weart, historian for the American Institute of Physics, have been framed by recurrent binary oppositions, particularly those of authority/victim, logic/feelings and nature/culture. 'On one side', he writes, 'stands the scientist with his dangerous devices, or the cruel parent, or the domineering male (especially government, industry, or military official), or the entire generalized threat from science and technology. On the other side stands the guinea pig, the rejected child, the enslaved worker, the dominated woman, the individual crushed by modern society'.[3] In the light of these comments, the rest of this chapter will explore two movies made by Hollywood in response to popular anti-nuclear protests in America.

Nuclear thriller: *The China Syndrome*

The China Syndrome (1978) was released at a time when the nuclear industry was in decline, due to spiralling costs as questions of safety led to regulatory delays.[4] The movie was a product of anti-nuclear advocacy amongst Hollywood creative personnel. Actors Jane Fonda and Jack Lemmon had both been long-time anti-nuclear campaigners: Lemmon had already narrated the voice-over for an anti-nuclear television documentary, *Plutonium: Element of Risk*, while Fonda had tried unsuccessfully to buy the movie rights to the Karen Silkwood story.

The plot of the movie begins with a safety problem at the (fictional) Ventana nuclear plant, which is followed by a negligent report by the Nuclear Regulatory Commission (NRC), rushed through in order to allow the go-ahead for a new power station. The NRC report misleadingly vindicates the existing safety procedures in the plant, blaming only minor faults for the problem, in a relay in the generator circuit and a stuck valve. As control room supervisor Jack Godell (Jack Lemmon) puts it at the time: 'the system works'. However, Jack soon discovers that the risks are more serious. Faulty welding has caused a radiation leak, and the construction company has falsified test X-rays to save money, so that some of the potentially faulty welds have not been checked. The company manager, when confronted by Jack about the cover-up, refuses further safety checks on financial grounds. Jack fears that, when the plant operates at full capacity, a build-up of water pressure may lead to meltdown, or 'The China Syndrome'. In the event, his predictions prove at least partly correct, as a meltdown is avoided only after he takes direct action and takes over the control room to rectify the situation, and is killed in the process by the police anti-terrorist squad.

The China Syndrome reinforces popular perceptions of nuclear science and technology as practices controlled by a corporate business system that places the demand for profit and expediency before human lives. The movie uses conventional images of dehumanizing machines controlled by a secret and deceptive elite to construct nuclear technology as unfamiliar, undemocratic, dangerous and alienating. In its characterization, then, the movie conforms to the conventions of melodrama. The conspiracy is run by unemotional scientists and unscrupulous businessmen, who wear neat dark suits and are one-dimensionally ruthless and corrupt. Conflict in the movie is Manichean, and resolved through heroic, vigilante action.

Human interest is provided by a dual narrative of political radicalization, as both Jack Godell and television news journalist Kimberly Wells (Jane Fonda) move from an initial position of naive acceptance of the industrial capitalist system to one of attempting to reform it from the inside. In doing so, Jack Godell becomes a self-sacrificial hero, dying to defend the plant that, significantly, he says he loves: his radicalization is, paradoxically, an act of loyalty to the nuclear industry. Jack's self-sacrifice inspires his colleague Ted Spindler (Wilford Brimley), an organization man with twenty-five years service, to follow his example in speaking out against corruption. Ryan and Kellner comment that 'the slow transformation of ordinary people into informed opponents of the corporate system probably appealed more to audiences than if the characters had begun as radicals'. Moreover, by providing the viewer with information about the conspiracy from the very beginning, the film, as they put it, 'operates by situating the audience in a position of privileged knowledge. The narrative suspense that comes into play when the corporation tries to prevent that knowledge from emerging into public light is thus linked to a political position'.[5]

Resistance in *The China Syndrome* is based on the gaining of information and knowledge which can be used as effective weapons against corporate corruption. The 16 mm film shot secretly in the nuclear plant by Richard Adams (Michael Douglas) is given as a definitive, objective version of the first nuclear accident, just as Kimberly's final live television broadcast reveals the truth about the second safety incident. The viewer is thus doubly reassured that audio-visual media technologies, when in the hands of men and women of honesty and goodwill, can give unproblematic access to truth. Moreover, deceptive representations, such as the falsified X-rays and the complacent lecture on the safety and efficiency of nuclear power given by the company public relations man Gibson (James Hampton) at the start of the film, are subsequently corrected in the light of truth and reason. Although information, knowledge and truth are at first controlled and suppressed by both the nuclear and the media industries, real, unbiased scientific expertise ultimately wins out. The movie thus represents dissident scientific experts as trustworthy authorities with access to objective truth. Technical advice for the movie came from three dissident nuclear engineers, Gregory C. Minor, Richard B. Hubbard and Dale Bridenbaugh (identified on screen as the consulting firm MHB Technical Associates), who had resigned their managerial jobs with General Electric in San Jose in 1976 amid great publicity, in protest over safety standards in the nuclear industry and the lack of public disclosure of information. Gregory Minor, a member of the Union of Concerned Scientists, has a cameo role in the movie as the nuclear expert who passes judgement on the risks posed by the Ventana nuclear plant. 'I may be wrong', he announces in what became a much-quoted speech, 'but I would say you're probably lucky to be alive. For that matter, I think we might say the same for the rest of southern California'. Although this rhetoric is inflated and apocalyptic, the rest of the movie does not contest it. Indeed, the cameo performance by a real-life scientist adds to the rhetoric of authenticity which the movie creates to disguise its melodramatic effects.

Kimberly's exposure of the truth about the nuclear plant on television at the end of the movie also vindicates the rationality and commitment to truth of investigative news journalism, which, when practised by people of integrity, is seen to inform the public about corrupt vested interests. Although the multiple screens in the television studio that open and close the movie suggest a confusing babble of competing voices and images, access to reliable truth through trustworthy mediation is nevertheless shown to be possible, and the competency of the mass media an effective guarantor of American democratic pluralism.

In constructing television as a vehicle for competent communication, and scientific expertise as an instrument of objective truth, *The China Syndrome* offers solutions which remain within the dominant paradigm of technocratic rationality, and therefore in keeping with the strategies of mainstream environmentalism. However, the movie is nuanced and ambiguous, and the solutions that it offers are tentative. Truth is signified as authentic not by the slick, scientific jargon of the industry men, but by the nervous inarticulateness of Jack's television broadcast, and the emotionality of both Ted Spindler's testimony and Kimberly's final live broadcast. Her final words ('I'm sorry I'm not very objective. Let's hope it doesn't end here') suggest a distrust of journalistic claims to detachment and objectivity, and only a tentative confidence in the possibility of reform. That the film ends without showing the response of the nuclear company to Kimberly's broadcast only compounds the openness of the ending. In one sense, that the meltdown is averted endorses the nuclear industry's faith in its safety policy of defence-in-depth. However, the escape does not resolve underlying anxieties concerning the future risks of technological breakdown.

The coincidental timing of the movie's release, twelve days before the nuclear accident at Three Mile Island, contributed to its topicality. Clarfield and Wiecek comment that the movie 'educated countless Americans in the jargon of high technology. The phenomenon of the film's title, core melts, the notorious phrase "an area the size of Pennsylvania", weld inspections, and other technical concepts that had previously been the domain of engineers and dedicated activists now became familiar to the people who had no opinions on nuclear power but who enjoyed a good story'.[6]

The release of the movie also generated heated debate in the American press on the subject of the ethics of cinematic representation. The *New York Times* invited six nuclear experts to comment on a film that they all agreed was released 'at a critical time in the brief history of man's attempt to harness nuclear power for peaceful purposes'.[7] Condemned by pro-nuclear scientists as scaremongering, the movie was defended by both the film-makers and their supporters in organizations such as the Union of Concerned Scientists.

Representatives of the nuclear industry pointed out what they saw as several inaccuracies and distortions in the movie's representation of their industry. Firstly, the movie ignores the issue of redundancy: nuclear plants do not have only one emergency core cooling system. Professor Norman Rasmussen indicated that the rupture of a large pipe, while possible, would only cause a problem 'if all of the many available back-up systems failed'. He added that this was 'highly unlikely, though not incredible'.[8]

Secondly, the time interval during which the plant approaches meltdown was greatly shortened in the movie for reasons of narrative economy. Eugene Cramer, of Southern California Edison, told the *Los Angeles Times* that a large meltdown would take at least sixty hours to enfold, allowing time for evacuation of surrounding areas. He added that the concrete shell around the reactor would retain most of the radioactivity.[9]

A third technical flaw in the film, from the point-of-view of the nuclear industry, was its suggestion that it would be difficult to shut down a reactor in the event of its control room being taken over by a terrorist or rebel worker such as Jack Godell. A shutdown, they pointed out, could be easily achieved from outside the control room.

John Taylor, vice-president of Westinghouse Electric, summed up a fourth response to *The China Syndrome*. 'What hurt me most about this film', he wrote, 'is that the utility chairman and the plant foreman are portrayed as morally corrupt and insensitive to their responsibilities to society. That view is inaccurate and incredible'. These criticisms demonstrate that the nuclear scientists interviewed by the *New York Times* were interpreting the movie in a literal, realist context, which implicitly rejected the defence of artistic licence. As viewers, they challenged the melodramatic conventions of the thriller genre, in particular its reliance on simplified characterization and the subordination of character to narrative action, on the grounds that such conventions led to an irresponsible misrepresentation of the nuclear industry as they knew it.

On the other hand, the movie was vigorously defended by the Union of Concerned Scientists, whose director, the economist Daniel Ford, described it as 'a major corrective to the myth that was drilled into us as children, that nuclear energy is a beautiful, endless, cheap source of electricity. The movie shows how that dream has been perverted by companies that operate the plant and how susceptible the program is to human error and industrial malfeasance'. The movie, he argued, is 'a composite of real events', providing 'a scenario that is completely plausible'.

Indeed, all of the problems with the nuclear reactor in *The China Syndrome* can be traced to actual safety problems experienced by American nuclear reactors since the 1950s. The shut down of the Commonwealth Edison Dresden II nuclear plant near Chicago on 5 June 1970 provided the screenwriters with two elements: the stuck pen recorder which misleads the reactor operators over the seriousness of the problem, and the added complication that the high pressure emergency cooling system is down for repairs and out of commission at the time of the accident. David Rossin, System Nuclear Research Engineer at Commonwealth Edison, commented that, 'There were enough obvious changes from what happened to make it an exciting story. If they had stuck to the facts, there wouldn't have been a movie'.[10]

The Browns Ferry reactor fire in Alabama on 22 March 1975 was the source of the low-water problem in the film's reactor. The fire, started negligently by an electrician, destroyed all five ECCS (emergency core cooling system) in one unit, demonstrating, for critics of nuclear power, the inadequacy of the defence-in-depth approach to reactor design. Clarfield and Wiecek describe the incident as pricking 'the conscience of many in industry and government about the lighthearted, indifferent attitudes towards safety that characterized the AEC's regulatory approach'. Although General Electric had declared it a 'non-accident' because nobody had been killed, the authors quote the reaction of Gregory Minor, script consultant on the movie, who had worked on the Browns Ferry reactor: 'to me it was a disaster. All the safety systems were gone! All of our backup systems were gone! I felt we were very, very lucky that we hadn't had a major catastrophe'.[11] As already noted, Minor subsequently resigned from General Electric in protest against safety standards.

The fake radiographs in the movie recall the Karen Silkwood affair. One of her former supervisors said that Silkwood had shown him falsified X-rays of fuel rods from the Kerr-McGee plant in Oklahoma. Jack Godell's hijacking of the control room recalls an incident in 1961 at a prototype military nuclear plant in Idaho, when three US servicemen engaged in manually lifting and re-engaging control rods were killed by a steam explosion. Gregory Minor produced an internal memo written by a Nuclear Regulatory Commission official which said that the accident had been caused on purpose by a disgruntled or psychotic employee.[12]

The China Syndrome was, then, to return to the words of the director of the Union of Concerned Scientists, a 'composite' of actual safety incidents involving nuclear reactors. However, his subsequent claim that the movie therefore provided a scenario that is 'completely plausible' is questionable. As journalist Robert Gillette pointed out, the movie fails to distinguish clearly between the *possibility* of nuclear accident and its *probability*. Such scientific niceties, it would seem, are irrelevant to the emotive requirements of the movie.[13] As Spencer Weart notes, popular fears of nuclear disaster exceed the statistical probabilities of harm. Yet to people 'who valued intuition and human feelings above logic,' he concludes, 'accident rates were just not the point. What really mattered was a question of values'.[14] The melodramatic form of *The China Syndrome*, then, though false to realistic plausibility, may be seen as true to the emotional experiences of those opposed to the nuclear power industry in the late 1970s.

Nuclear biopic: *Silkwood*

Like *The China Syndrome*, *Silkwood* represents the nuclear industry as more concerned with production targets than with worker safety. Company officials are similarly complacent in their attitude to the dangers of nuclear radiation. The manager of the Kerr-McGee plant is first seen looking down on his workers from an observation platform, a figure of detached, elitist surveillance.

Moreover, the authority of official industry 'experts' is again questioned, in that the scientific estimates of safety given by the company, the government and the medical profession are all contested at some point in the narrative. According to Thelma (Sudie Bond), the company doctor 'trained as a vet'. When Karen (Meryl Streep) is contaminated, her partner Drew (Kurt Russell) repeats this sceptical attitude. 'Doctors are all goddam liars', he complains.

Silkwood differs from *The China Syndrome*, however, in its emphasis on issues of unionization and workers' rights to health and safety, within an explicit context of gender and class relationships. The movie also carefully observes the ways in which the workers in the nuclear plant rationalize their political quiescence, as their dependency on the company for job security leads them to form defence mechanisms, particularly the use of black humour to offset their sense of powerlessness. Radiation, jokes Dolly (Cher), 'makes your nipples turn green'. Karen Silkwood begins the movie with a similarly flippant attitude, but begins to reject such bad faith, and realizes the need to resist the status quo when her friend Thelma is accidentally contaminated with radiation. After this scene, the movie cuts to Karen cleaning her kitchen, and confronting Dolly with the mouldy spaghetti she has left in the refrigerator. The sense of invasive contamination by invisible, unfamiliar sources is thereby extended from nuclear radiation to other forms of disease and pollution. Karen's obsessive cleaning is a psychological response that is a concrete way of reasserting a feeling of personal control over the perceived threat to her everyday sense of identity. *Silkwood* goes on to explore the vicissitudes of human relationships as the lived context in which technological risks are experienced, negotiated and survived. Nuclear hazards affect the inner lives of the characters, and also invade the body's privacy, a violation most graphic-ally expressed in the scenes where potentially contaminated bodies, vulnerable and fragile, are subjected to violent scrubbing by white-coated technicians in the company showers.

In particular, the narrative concentrates on the effect that Karen's politicization has on her relationship with Drew, who finds it difficult to come to terms with the time she spends working for the union. Drew's sense of powerlessness and his retreats into alcohol suggest the inadequacy of traditional conceptions of masculine heroism to deal with the complexities of nuclear risk. 'Don't give me a problem I can't solve', he tells Karen. In its exploration of the relationship between Karen and Drew, the movie breaks away from the Manichean mode of characterization typical of melodrama, and develops instead in a more naturalistic mode, producing graduated characters in situations of relative psychological complexity. Nevertheless, the narrative of Karen's politicization is part of the concern for heroism typical of the Hollywood biopic, while the emphasis on the personal growth of its female protagonist also draws on the conventions of the 'woman's picture'.

Resistance, as in *The China Syndrome*, is based on instinct and emotion, combined with the scientific knowledge provided by the counter-expertise of the union scientists. When Karen is contaminated with radiation, she reads the union handbook for the first time. 'It says here, plutonium gives you cancer. It says it flat out', she tells Dolly. Scientific information thus provides the rational grounds for Karen's actions, by validating her own emotional experience. Yet emotional truths again take the place of the complexities and uncertainties of risk analysis.

That *Silkwood* is ultimately evasive in its representation of the nuclear industry is perhaps unsurprising given its source in near contemporaneous historical events. Thus although the movie directly implicates an employee of Kerr-McGee in malpractice, when technician Winston tampers with the X-rays of faulty fuel rods being sent to a new breeder reactor testing at Hanford, it carefully stops short of implying that the corporate bosses are also involved in the fraud.

Moreover, the movie is careful to treat nuclear issues in an even-handed way. The labour union is not idealized, but is shown to be mainly interested in using the health and safety issues addressed by Silkwood as a lever in its contract negotiations with the company. Audience sympathy for union leader Paul Stone (Ron Silver) is lost when he is unavailable to give Karen the support she needs when putting herself in danger by spying on the company. Even Winston (Craig T. Nelson), the company man who falsifies the X-rays, has some justification in accusing Stone of bandwagon-jumping, asking him: 'If you're so worried about us, where the hell were you in the beginning?'

Such political caution crucially shapes the ending of the movie. The film-makers were careful to acknowledge the factual uncertainties and alternative explanations of Silkwood's death, and to qualify her supporters' allegation that she was murdered by a company official. When Karen dies in the car crash at the end of the film, the camera pans ambiguously past Thelma, the newly politicized activist, to Dolly, betrayer of her friend and guilty collaborator with the company. Yet the actual car crash is not shown: there is a cut from Karen driving with another car's headlights visible behind her, to her car in the ditch: a narrative ellipsis that acknowledges the uncertainties in the historical evidence. The car behind may have shunted her off the road, as her supporters believed, but the action is not shown. The movie is thus careful not to implicate Kerr-McGee unequivocally in Silkwood's death, while the company's defence, that her car accident was caused by her own careless driving, possibly under the influence of drugs, is not refuted. Indeed, the Afterword gives this possible explanation equal emphasis, the narrative having already established that Silkwood was a careless driver, a taker of medication and a user of marihuana. As Karen's earlier rendition of 'Amazing Grace' is replayed on the soundtrack, the final scenes of the movie displace political issues into the therapeutic. Politically, then, *Silkwood* stops short of constructing its heroine as a victim of a nuclear police state. Instead, her life is turned into an American myth of redemption and personal growth.[15]

Nuclear Accidents and Representation

Historian Hayden White includes 'pollution of the ecosphere by nuclear explosions and the indiscriminate disposal of contaminants' in his list of what he calls those 'modernist events' which are 'anomalous' in their 'resistance to inherited categories and conventions for assigning them meanings'. The modern media, he continues, represent these events 'in such a way as to render them, not only impervious to every effort to explain them but also resistant to any attempt to represent them in a story form'. Moreover,

> not only are *modern* post-industrial 'accidents' more incomprehensible than anything earlier generations could possibly have imagined (think of Chernobyl), the photo and video documentation of such accidents is so full that it is difficult to work up the documentation of any one of them as elements of a single 'objective' story. Moreover, in many instances, the documentation of such events is so manipulable as to discourage the effort to derive explanations of the occurrences of which the documentation is supposed to be a recorded image. 'It is no accident,' then, that discussions of the modernist event tend in the direction of an aesthetics of the sublime-and-the-disgusting rather than that of the beautiful-and-the-ugly.[16]

The exploration of the cinematic representation of nuclear accidents undertaken in this chapter provides evidence both to support White's assertions and to qualify them. An 'aesthetics of the sublime-and-the-disgusting' may account successfully for the representation of nuclear issues in horror or science fiction movies, genres that recent Hollywood cinema has revived in order to displace anxieties about nuclear accidents into symbolic images of alien invasion. The Toxic Avenger was born out of a vat of nuclear waste to become the scourge of corrupt, polluting industry, in the three *Toxic Avenger* movies (1985–9). The Hollywood version of *Godzilla* (1998), based on the 1950s Japanese comic-book monster and 1960s movie cycle, updated the story by blaming fallout from the nuclear explosion at Chernobyl for causing the lizard to mutate genetically into a monster that threatens New York City. Such images of sublimity and disgust are appropriate metaphors for the sense of incomprehensibility evoked by the scale of nuclear accidents, and the perceived powerlessness of individuals to do anything about them.

On the other hand, the notion that the nuclear age produces a crisis in 'story form', asserted by White in the quotation above, appears to fit less convincingly the movies discussed in this chapter. At the very least, the crisis does not appear to have been recognized by Hollywood: if traditional, linear narrative forms are inadequate to the task of representing nuclear events, then this is a message that has failed to register with either the film-makers or the popular audiences for their films. Admittedly, in French cinema, Alain Renais's *Hiroshima, Mon Amour* (1959) made explicit the link between the dropping of the atomic bomb and the collapse of the possibility of stable meaning and linear narrative structure: a modernist text to match White's 'modernist' event. Yet the complex, non-linear narrative structure of this European art film is in marked contrast to the narrative techniques familiar to Hollywood's commercial aesthetic. In comparison, both movies discussed in detail in this chapter continue to place confidence in the efficacy of traditional, mimetic techniques of representation (the 16 mm film and television broadcast in *The China Syndrome*, and the X-rays showing industrial negligence in *Silkwood*), as well as their own traditional forms of linear narration, in a way that should qualify White's hyperbolic insistence that such confidence is no longer possible. Indeed, however 'manipulable', in White's words, the photographic and filmic documentation of nuclear accidents is shown to be in these movies, they both nevertheless ultimately endorse the efficacy of such media to narrate 'a single "objective" story' about the nuclear power industry in the United States. Moreover, they do so in narratives that, while conforming closely to generic conventions, are sufficiently adaptable to leave their endings relatively open and unresolved, thereby rejecting forms of closure that might appear inadequate to the subject being represented. Nuclear accidents, then, like all of the environmental issues discussed in this book, appear to be no more 'resistant to any attempt to represent them in a story form', in White's words, than any other events in the real world. Indeed, the melodramatic mode, to return to the words of Linda Williams quoted in the Introduction to this book, continues to provide Hollywood cinema with the aesthetic means for the 'dramatic revelation of moral and emotional truths through a dialectic of pathos and action'.[17] That the 'moral and emotional truths' about nuclear power revealed by the movies discussed in this chapter were openly contested in their public reception suggests not that the representational forms used to communicate these truths are inadequate to the task, but, on the contrary, that the play between melodrama and realism in Hollywood cinema continues to be an artistically productive way of dramatizing the events of the nuclear age.

Conclusion

The topical and urgent environmental concerns around which this book is structured have been mediated by Hollywood's commercial aesthetic to produce a wide range of popular films across many different genres. The movies have demonstrated that 'oblique but unbroken connection to the historical world' described by Michael Wood, a complex relation based, as he puts it, on 'wish, echo, transposition, displacement, inversion, compensation, reinforcement, example, warning'.[1]

Yet, as Richard Maltby and Ian Craven rightly conclude, a proper study of Hollywood movies will 'look less to the discovery of profound meanings or concealed purposes in its texts, and more to the equally difficult task of articulating the silences and equivocations, the plenitudes, excesses, and banalities of their surfaces'.[2] *Green Screen* will therefore conclude with three key moments that demonstrate the mythic power, banalities and ideological limitations of Hollywood cinema's embrace of environmentalist advocacy.

In *Dances With Wolves* (1990), army deserter John Dunbar (Kevin Costner) finds what he calls a 'trusted friend' in a wolf, whom he names Two Socks. Eventually, in an epiphanic moment of mutual trust between human and wild animal, Two Socks eats from Dunbar's hand, in a close-up accompanied by sentimental pastoral string music. The white man, formerly complicit in the conquest of the wilderness and its inhabitants, is redeemed by his contact with this representative of benign nature. When the Sioux Chief renames him 'Dances With Wolves', his redemption is complete, and the white man is reborn in harmony with nature, like the wild wolf and the Sioux.

This scene typifies the therapeutic tendency in Hollywood's approach to environmental concerns, as the notion of the deep ecological Native American combines with that of the benevolent wolf to signify the healing of traumatic divisions between white man and Indian, and human being and animal. What is repressed in the process are two aspects central to Parts One and Two of this book: a recognition of Native Americans as members of complex, heterogeneous and historical cultures, and of wild animals as more than friendly and subordinate companions to human beings. Such considerations play no part in the conservationist advocacy of this film, which emerges instead from a need to renew the hegemony of the white American male, restored to innocence through mythic contact with the redemptive purity of nature.[3]

A second scene dramatizes the sometimes contradictory attitude to technology in Hollywood environmentalist movies. The opening sequence of the science fiction film *Silent Running* (1971) shows close-ups of yellow flowers, a snail, a turtle, strawberries, and a frog. But by the year 2000 nature has been eradicated on Earth, and exists only as a simulated environment in a spaceship, in the hope that reforestation can one day take place when the Earth's atmosphere is less polluted. The plot of the movie turns on the government's decision to abandon the reforestation project for financial reasons.

Popular resistance in the movie, as in many environmentalist films, takes the form of the individual vigilante, Freeman Lowell (Bruce Dern), a St Francis-figure dressed in a monkish white habit, who talks to the animals in the forest, and uses justified violence to fight for his own inner sense of justice. Lowell is also the archetypal Jeffersonian farmer tending his garden in the wilderness, and the inner-directed individualist fighting the totally administered society of the bureaucratic, corporate state. 'You're a hell of an American', his astronaut colleague says mockingly. Lowell replies: 'I think I am'.

Silent Running is typical of many of the environmentalist movies discussed in this book in its uneasy combination of nostalgia for a seemingly lost authentic relationship between human beings and non-human nature, before the despoilations of modernity, with a reliance on a technological fix to solve environmental problems. Paradoxically, then, Lowell eventually saves authentic nature by using technology and artifice. When he discovers that the forest in the spaceship's dome is dying though lack of sunlight, he deploys artificial lights to 'do the job that the sun would do'. Finally, when ordered by the government on Earth to jettison the dome, he entrusts the future conservation of the forest to a humanized robot named Dewey. The closing images of *Silent Running* show Dewey tending the forest with a watering can, as Lowell blows himself up with the spaceship. Yet the technological image of the robot ironically contradicts the folksy, hippie authenticity of the film's closing song, 'Rejoice in the Sun' sung by Joan Baez, which evokes a romantic view of nature as pristine wilderness: 'Fields of children running wild in the sun. Like a forest is your child, growing wild in the sun'. Wild nature has been restored, paradoxically, by the interventions of a benign technology. The robot to whom the care of nature is entrusted is a convivial technology, childlike and endearingly clumsy. Machines may have become intelligent and sentient, but are also fantasized as benign, unthreatening and reassuringly humanized. Nevertheless, the ending of *Silent Running* remains open and uncertain. The forest in the dome, says Lowell, is like a note in a bottle: one day someone may find it.

The appeal made by *Silent Running* to new technologies as a solution to environmentalist problems recalls other movies discussed in this book, such as *On Deadly Ground* and *Free Willy 2*. The technological fix provides an apolitical solution to environmental problems, and in this sense endorses the values of mainstream environmentalism. Yet the optimism of these movies remains muted, and the narrative resolution demanded by the melodramatic mode does not close off a sense that environmental problems are ongoing, and not totally amenable to such fixes.

Anxieties over the death of non-human nature raised by *Silent Running* have become the concern not only of American preservationists such as Bill McKibben, but also of postmodernist critics such as Frederic Jameson. 'True, nature is resistant, and infinite in its depth', wrote Henri Lefebvre, 'but it has been defeated, and now waits only for its ultimate voidance and destruction'.[4] This book, on the other hand, has refused to jump to such conclusions in the absence of convincing evidence to support them, and has consequently attempted to steer a path clear of such fatalism and hyperbole, on the grounds that they are not only philosophically dubious, but also limiting with regard to an effective environmental politics. Instead, this book has assumed that although Hollywood environmentalist movies construct simulations of nature, non-human nature nevertheless continues to exist as a referent external to the discourses human beings invent to describe or explain it. That shots of orcas in the *Free Willy* movies are often produced by means of mechanical Animatronic replicas does not mean that real orcas no longer exist, whether in real marine parks or in the world's oceans. Such a position is one of critical, but not naïve, realism: it does not deny that those real orcas are constructed in human discourses for a range of symbolic purposes, as the popularity of both whale-watching and the Free Keiko campaign demonstrates. It has been the task of this book to analyse the variety of such discursive practices as they are constructed in Hollywood movies about the natural and built environments.

The movies discussed in this book remain a small part of the overall output of a Hollywood film industry whose environmental sensibilities are always likely to be moderated by its vested interest in promoting commodity consumption as a social good. Even movies with self-consciously 'green' messages thus tend to avoid questioning the central place that consumerism has in American society. Consumerism is, of course, second nature in Hollywood movies, and the 'teen' movie genre, given a central place in contemporary Hollywood production due to the industry's commercial need to make movies that appeal to a young audience, is so complicit in consumerist values that self-questioning has been restricted to the evasive irony of *Wayne's World* (1992), which pokes fun at its own product-placement strategies. This impasse is further illustrated by the final exemplary film, *Bio-Dome* (1996), which tries to reconcile concern for the environment with an ideology of consumption-as-usual.

When Generation X'ers Bud (Pauly Shore) and Doyle (Stephen Baldwin) are accidentally trapped inside the Arizona Bio-Dome, having mistaken it for a shopping mall, they hold a party which damages the 'harmony' and 'balance' of the artificial ecosystem. Converted to environmental awareness by their sensible girlfriends, Monique (Joey Adams) and Jen (Teresa Hill), the boys then resolve to restore the dome's 'homeostasis'. So the slacker generation faces up to its moral responsibilities, and the mad scientist Dr Noah Faulkner (William Atherton) is defeated.

Nevertheless, the environmentalist message of the movie is contradicted by its assumption, common to the teen genre, that consumerism is second nature. At the end of the film, the boys' heroic efforts at restoring the Bio-Dome win them a contract to endorse a new soft drink, the Colonic Cannon. Moreover, when the girls drive to an Earth Day meeting in their shiny red sports car, the possible contradiction is not commented upon. Instead, American commodity consumption, typified by car usage, continues to be naturalized in popular cinema. The movie's optimistic, 'win-win' solution reassures its audience that it can both 'save the planet' and continue to consume at current levels. *Bio-Dome* is thus able to exploit the topicality of its environmentalist concerns, while endorsing business-as-usual for American consumer capitalism. It is within such contradictions, compromises and evasions that the Hollywood movies discussed in this book perform their cultural work.

NOTES

Preface

1 'Industry Set for Green Films of Fall', *Hollywood Reporter*, 6 September 1990, p. 21.
Terry George, 'Hollywood Goes Green', *Audubon*, March/April 1992, p. 86.
Most of the movies written about in this book were made in the era of the 'New Hollywood', a shorthand term for the global multi-media entertainment conglomerations that have dominated the film industry since the 1980s. The term 'Hollywood' thus includes both movies that were made by the Hollywood studios themselves, and independent productions either co-financed or distributed by the major companies.

2 Neil Smith, 'The Production of Nature', in *FutureNatural: Nature, Science, Culture*, ed. George Robertson, Melinda Mash, Lisa Tickner, Jon Bird, Barry Curtis and Tim Putnam (London and New York: Routledge, 1996), p. 42.

3 Richard Maltby and Ian Craven, *Hollywood Cinema* (Oxford and Cambridge Mass.: Blackwell, 1995), pp. 383, 390.

4 Stephen Prince, *Visions of Empire: Political Imagery in Contemporary American Film* (New York: Praeger, 1992), p. 40.

5 Kate Soper, 'Nature/nature', in *FutureNatural*, ed. Robertson *et al.*, p. 30. For an influential statement of the extreme social constructionist position on both nature and science, see Donna J. Haraway, 'Situated Knowledges: The Science Question in Feminism and the Privilege of Partial Perspective', in *Simians, Cyborgs, and Women: The Reinvention of Nature* (New York: Routledge, 1991), pp. 183–201.

6 Ibid., p. 31. See also Kate Soper, *What is Nature? Culture, Politics and the Non-Human* (Oxford and Cambridge Mass.: Blackwell, 1995); Christopher Norris, *Against Relativism: Philosophy of Science, Deconstruction and Critical Theory* (Oxford: Blackwell, 1997). On anti-science sentiment in ecocriticism, see Glen Love, 'Science, Anti-Science, and Ecocriticism', *ISLE: Interdisciplinary Studies in Literature and the Environment*, 6.1 (Winter 1999), 65–81; Ursula K. Heise, 'What is Ecocriticism?', Association for the Study of Literature and the Environment web-site, at http://www.asle.umn.edu/

Introduction

[1] Cameron Stauth, 'Eco Trip: Hollywood Turns Green-eyed in a Multipicture Deal with Planet Earth', *American Film*, November 1990, p. 16.

[2] Barbara Adam, *Timescapes of Modernity: The Environment and Invisible Hazards* (London and New York: Routledge, 1998), p. 17.

[3] Ella Shohat and Robert Stam, *Unthinking Eurocentrism: Multiculturalism and the Media* (London and New York: Routledge, 1994), p. 182.

[4] Linda Williams, 'Melodrama Revised', in *Refiguring American Film Genres: Theory and History*, ed. Nick Browne (Berkeley, Los Angeles and London: University of California Press, 1998), pp. 42, 54.

[5] Ibid., pp. 53, 48.

[6] Leo Braudy, 'The Genre of Nature: Ceremonies of Innocence', in *Refiguring American Film Genres: Theory and History*, ed. Browne, p. 290.

[7] Robert Ray, *A Certain Tendency of the Hollywood Cinema, 1930–1980* (Princeton NJ: Princeton University Press, 1985), p. 253.

[8] Williams, 'Melodrama Revised', p. 80.

[9] Greg Myers, '"The Power is Yours": Agency and Plot in *Captain Planet*', in *In Front of the Children: Screen Entertainment and Young Audiences*, ed. Cary Bazalgette and David Buckingham (London: BFI Publishing, 1995), p. 73.

[10] Kate Soper, *What is Nature? Culture, Politics and the Non-Human* (Oxford and Cambridge Mass.: Blackwell, 1995), p. 262.

[11] Michael Ryan and Douglas Kellner, *Camera Politica: The Politics and Ideology of Contemporary Hollywood Film* (Bloomington and Indianapolis: Indiana University Press, 1990), p. 105.

[12] Ibid., p. 104.

[13] Richard Slotkin, *Gunfighter Nation: The Myth of the Frontier in Twentieth-Century America* (New York: Atheneum, 1992), pp. 13–14.

[14] Ibid., p. 23.

[15] See *Realism and the Cinema: A Reader*, ed. Christopher Williams (London: Routledge & Kegan Paul, 1980); Terry Lovell, *Pictures of Reality: Aesthetics, Politics and Pleasure* (London: BFI Publishing, 1983). As a mode of European cinema, 'modernism', in the sense of a cinema of formal, self-reflexive experiment, is less relevant to a study of the commercial aesthetics of Hollywood cinema. See David Bordwell, *On the History of Film Style* (Cambridge Mass. and London: Harvard University Press, 1997), ch. 4.

[16] Quoted in Bernard Weinraub, 'Good Guys, Bad Guys', *International Herald Tribune*, 13 January 1999, p. 9.

[17] Jonathan Harr, *A Civil Action* (New York: Vintage Books, 1995). For an alternative account of the case, critical of Harr, see Dan Kennedy, 'Take Two', *Boston Phoenix*, 2 January 1998, pp. 14–16; for criticism of the suppression of the families' viewpoints in both book and film, see Dan Jewel and Mark Dagostino, 'A Civil Warrior', *People*, 8 February 1999, pp. 79–82 and David Tyler, 'Real Story is Lost in Movie Version of 'A Civil Action', *Daily Transcript* (Dedham, Massachusetts), 1 June 1999, p. 9.

[18] For the distinction between 'alignment' and 'allegiance', see Murray Smith,

Engaging Characters: Fiction, Emotion and the Cinema (Oxford: Oxford University Press, 1995), pp. 187–8. On the 'reality effect' of realist fictions, see Roland Barthes, 'The Reality Effect', in *The Rustle of Language*, trans. Richard Howard (Oxford: Blackwell, 1986), pp. 141–8.

[19] Noël Carroll, *Theorizing the Moving Image* (Cambridge: Cambridge University Press, 1996), p. 244.

[20] David Bordwell and Kristin Thompson, *Film Art: An Introduction* (New York and London: McGraw-Hill, Inc., 1993), p. 180.

[21] On the distinction between 'graduated' characters, who represent 'a spectrum of moral gradations rather than a binary opposition of values', and Manichean characters, who embody an 'unqualified opposition of good and evil values', see Smith, *Engaging Characters*, pp. 197, 297.

[22] Williams, 'Melodrama Revised', p. 42.

[23] Steve Neale, '"You've Got To Be Fucking Kidding!" Knowledge, Belief and Judgement in Science Fiction', in *Alien Zone: Cultural Theory and Contemporary Science Fiction*, ed. Annette Kuhn (London: Verso, 1990), p. 164.

[24] Anonymous review of *Day of the Animals*, *Variety*, 8 June 1977, p. 26.

[25] Stephen Neale, 'The Same Old Story: Stereotypes and Difference', *Screen Education*, 31–2 (Autumn/Winter 1979–80), pp. 33–7.

[26] On the 'revenge of Nature' in horror movies, see Robin Wood, 'An Introduction to the American Horror Film', in *Movies and Methods Vol. II*, ed. Bill Nichols (Berkeley, Los Angeles and London: University of California Press, 1985), p. 207.

[27] Maurice Yacomar, 'The Bug in the Rug: Notes on the Disaster Genre', in *Film Genre Reader II*, ed. Barry K. Grant (Austin: University of Texas Press, 1995), p. 262.

[28] Tom Athanasiou, *Slow Reckoning: The Ecology of a Divided Planet* (London: Secker and Warburg, 1997), pp. 74–6.

[29] M. Jimmie Killingsworth and Jacqueline S. Palmer, 'Millennial Ecology: The Apocalyptic Narrative from *Silent Spring* to *Global Warming*', in *Green Culture: Environmental Rhetoric in Contemporary America*, ed. Carl G. Herndl and Stuart C. Brown (Madison and London: University of Wisconsin Press, 1996), p. 41.

[30] Richard Maltby and Ian Craven, *Hollywood Cinema* (Oxford and Cambridge Mass.: Blackwell, 1995), p. 281.

[31] Michael Wood, *America in the Movies, or 'Santa Maria, It Had Slipped My Mind'* (New York: Columbia University Press, 1975), p. 18.

[32] Stephen Prince, 'True Lies: Perceptual Realism, Digital Images, and Film Theory', in *Film Quarterly: Forty Years—A Selection*, ed. Brian Henderson and Ann Martin (Berkeley, Los Angeles and London: University of California Press, 1999), p. 400.

Chapter One

[1] See Peter Coates, *In Nature's Defence: Americans and Conservation* (Keele: British Association for American Studies, 1993); H.P. Caulfield, 'The Conservation and Environmental Movements: An Historical Analysis', in *Environmental Politics and Policy: Theories and Evidence*, ed. J.P. Lester (Durham and London: Duke University Press, 1989).

[2] For various critiques of mainstream environmentalism, see Andrew Ross, 'Introduction', *The Chicago Gangster Theory of Life* (London and New York: Verso, 1994); Ariel Salleh, *Ecofeminism as Politics: Nature, Marx and the Postmodern* (London and New York: Zed Books, 1997); Arran E. Gare, *Postmodernism and the Environmental Crisis* (London and New York: Routledge, 1995).

[3] Tom Athanasiou, *Slow Reckoning: The Ecology of a Divided Planet* (London: Secker and Warburg, 1996), p. 6.

[4] David Harvey, *Justice, Nature and the Geography of Difference* (Oxford and Cambridge Mass.: Blackwell, 1996), pp. 174–5.

[5] Martin W. Lewis, *Green Delusions: An Environmentalist Critique of Radical Environmentalism* (Durham and London: Duke University Press, 1992). See also Paul R. Gross and Norman Levitt, *Higher Superstition: The Academic Left and Its Quarrels with Science* (Baltimore: Johns Hopkins University Press, 1994); Martin W. Lewis, 'Radical Environmental Philosophy and the Assault on Reason', in *The Flight From Science and Reason*, ed. Paul R. Gross, Norman Levitt and Martin W. Lewis (New York: New York Academy of Sciences, 1996).

[6] Robert Ray, *A Certain Tendency of the Hollywood Cinema, 1930–1980* (Princeton NJ: Princeton University Press, 1985), p. 37. See also Ronald Davis, 'Paradise among the Monuments: John Ford's Vision of the American West', *Montana: The Magazine of Western History*, Summer 1995, pp. 49–62.

[7] Peter B. Kyne, *The Valley of the Giants* (New York: Grosset and Dunlap, 1918). I have been unable to locate a print of the 1919 movie.

[8] Simon Schama, *Landscape and Memory* (London: HarperCollins, 1995), p. 188.

[9] Nancy K. Anderson, '"The Kiss of Enterprise": The Western Landscape as Symbol and Resource', in *The West as America: Reinterpreting Images of the Frontier, 1820–1920*, ed. William H. Truettner (Washington and London: Smithsonian Institution Press, 1991), pp. 268–81. See also Susan Schrepfer, *The Fight to Save the Redwoods: A History of Environmental Reform 1917–1978* (Wisconsin and London: University of Wisconsin Press, 1983), p. 7.

[10] See, for example, 'Forest Monarchs' and 'California Redwoods' (*c.* 1875), in Gordon Hendricks, *Albert Bierstadt: Painter of the American West* (New York: Harry N. Abrams, Inc., 1974), p. 237 and 'Grisly Giant 86 feet in circumference 225 feet high Mariposa Grove California 1861', in Peter E. Palmquist, *Carelton E. Watkins: Photographer of the American West* (Albuquerque: University of New Mexico Press, 1983), p. 30, plate 10.

[11] On forestry policy and conservationism, see Alfred Runte, *National Parks: The American Experience* (Lincoln: University of Nebraska Press, 1979), pp. 33–64; Hans Huth, *Nature and the American: Three Centuries of Changing Attitudes* (Berkeley and Los Angeles: University of California Press, 1957), pp. 142ff. and 192ff.;

Patricia Nelson Limerick, *The Legacy of Conquest: The Unbroken Past of the American West* (New York and London: Norton, 1987), pp. 295–9. The western *Riders of the Whistling Pines* (1949), in which Gene Autry supervises the spraying of a moth-infested forest with 'safe' levels of DDT, also endorses the conservationist forestry policies of its day.

12 Stephen Budiansky, *Nature's Keepers: The New Science of Nature Management* (London: Phoenix Giant, 1996), pp. 7, 5–6.

13 Daniel B. Botkin, *Discordant Harmonies: A New Ecology for the Twenty-First Century* (New York and Oxford: Oxford University Press, 1990).

14 Budiansky, *Nature's Keepers*, p. 11.

15 William Cronon, 'The Trouble with Wilderness; or, Getting Back to the Wrong Nature', in *Uncommon Ground: Rethinking the Human Place in Nature*, ed. William Cronon (New York and London: W.W. Norton & Company, 1996), pp. 81–2.

16 Matt Cartmill, *A View to a Death in the Morning: Hunting and Nature Through History* (London and Cambridge Mass.: Harvard University Press, 1993); Ralph H. Lutts, 'The Trouble With Bambi: Walt Disney's *Bambi* and the American Vision of Nature', *Forest and Conservation History*, 36.4 (October 1992), pp. 160–71; Gregg Mitman, *Reel Nature: America's Romance with Wildlife on Film* (Cambridge Mass. and London: Harvard University Press, 1999), pp. 111–12.

17 Stephen Pyne, *Fire in America: A Cultural History of Wildland and Rural Fire* (Princeton NJ: Princeton University Press, 1982), p. 194.

18 Ibid., p. 176.

19 Limerick, *The Legacy of Conquest*, p. 305.

20 Paul Schullery, '*Bambi* and the fires of Yellowstone', *Backpacker*, December 1990, pp. 95–6.

21 See David Payne, '*Bambi*', in *From Mouse to Mermaid: The Politics of Film, Gender, and Culture*, ed. Elizabeth Bell, Lynda Haas and Laura Sells (Bloomington and Indianapolis: Indiana University Press, 1995), pp. 137–147; Patrick D. Murphy, '"The Whole Wide World Scrubbed Clean": The Androcentric Animation of Denatured Disney', in ibid., pp. 125–30.

22 See Anne Taylor Fleming, 'Turning Stars Into Environmentalists', *New York Times*, 25 October 1989, C8; Jane Lieberman, 'Hollywood Who's Who Backs EMA', *Variety*, 10 August 1992, p. 50; Joan Goodman, 'How Hollywood Whispers in the President's Ear', *Sunday Times Magazine*, 23 March 1997, pp. 20–4; Phillip W.D. Martin, 'Quiet, Please, on the Set: Birds Nesting', *New York Times*, 19 February 1995, pp. 13–15.

Hollywood environmental advocacy was further reflected at this time in *Heaven is Under Our Feet*, ed. Don Henley and Dave Marsh (New York: Berkley Books, 1991), a collection of writings in aid of the Walden Woods Project, founded by rock musician Don Henley to save Walden Woods from development. The collection includes appeals by movies stars Tom Cruise, Ted Danson, Robert Redford, Whoopi Goldberg, James Earl Jones and Meryl Streep, amongst many others.

23 Environmental Media Association web-site, at http://www.epg.org/BG.ema.html.

24 Michael Carlton, 'Disney Environmentalists', *Southern Living*, 30.12 (December

1995), p. 90.

[25] John Sterling, 'The World According to Disney', *Earth Island Journal*, Summer 1994, pp. 32–3. Michael Eisner defends his plans for the theme park in *Work in Progress* (New York: Random House, 1998).

[26] Andrew Ross, 'Introduction', *No Sweat: Fashion, Free Trade, and the Rights of Garment Workers*, ed. Andrew Ross (New York and London: Verso, 1997), p. 9. For The National Labor Committee's 'Open Letter to Walt Disney', ibid., pp. 95–112. Gregg Easterbrook has a different set of figures, but makes the same point with them, in *A Moment on the Earth: The Coming Age of Environmental Optimism* (Harmondsworth: Penguin, 1995), p. 677.

[27] Alexander Wilson, *The Culture of Nature: North American Landscape from Disney to the Exxon Valdez* (Oxford: Blackwell, 1992), p. 118. Mitman analyses Disney's True-Life Adventure film *Nature's Half-Acre* (1951) in terms of Linnaean notions of the balance of nature, in *Reel Nature*, p. 128.

[28] On the concept of 'place' in Darwin, see Donald Worster, *Nature's Economy: A History of Ecological Ideas* (Cambridge: Cambridge University Press, 1994), p. 157. The metaphor of the circle also occurs in Barry Commoner, *The Closing Circle: Nature, Man and Technology* (New York: Alfred A. Knopf, 1972).

[29] Terry Diggs, 'Welcome to the All-Republican Cineplex', *Legal Times*, 19 June 1995, p. 58.

[30] Worster, *Nature's Economy*, p. 296.

[31] Ted Kerasote, 'Disney's New Nature Myth', *Audubon*, November/December 1994, p. 132.

[32] Stan Steiner, *The Vanishing White Man* (Norman and London: University of Oklahoma Press, 1987), p. 113. Also quoted in Bill Devall and George Sessions, *Deep Ecology: Living as if Nature Mattered* (Salt Lake City: Gibbs Smith, Publisher, 1985), p. 97.

[33] Sallie Hofmeister, 'In the Realm of Marketing, the "Lion King" Rules', *New York Times*, 12 July 1994, p. D17.

[34] Worster, *Nature's Economy*, p. 313.

Chapter Two

[1] Andrew Ross, *The Chicago Gangster Theory of Life* (London: Verso, 1994), p. 205. See also Max Oelschlaeger, *The Idea of Wilderness: From Prehistory to the Age of Ecology* (New Haven and London: Yale University Press, 1991); Roderick Nash, *Wilderness and the American Mind* (New Haven and London: Yale University Press, 1982).

[2] The Wilderness Act 1964, quoted in Patricia Nelson Limerick, *The Legacy of Conquest: The Unbroken Past of the American West* (New York and London: Norton, 1987), p. 309.

[3] Bill McKibben, *The End of Nature* (New York: Random House, 1989); Thomas Berry, *The Dream of the Earth* (San Francisco: Sierra Club Books, 1988).

[4] William Cronon, in *Uncommon Ground: Rethinking the Human place in Nature* (New York and London: W.W. Norton & Company, 1996), pp. 83, 85.

[5] Neil Smith, *Uneven Development: Nature, Capital and the Production of Space* (Oxford

and Cambridge Mass.: Basil Blackwell, 1990), p. 16.

6 John Urry, *The Tourist Gaze: Leisure and Travel in Contemporary Societies* (London: Sage, 1992), p. 98; Peter J. Schmitt, *Back to Nature: The Arcadian Myth in Urban America* (Baltimore and London: Johns Hopkins University Press, 1990), p. 147.

7 On the aesthetics of nature photography, see Deborah Bright, 'The Machine in the Garden Revisited: American Environmentalism and Photographic Aesthetics', *Art Journal*, 51.2 (Summer 1992), pp. 60–71.

8 David Bordwell, Janet Staiger and Kristin Thompson, *The Classical Hollywood Cinema: Film Style and Mode of Production to 1960* (New York: Columbia University Press, 1985), p. 345.

9 Barbara Novak, *Nature and Culture: American Landscape Painting, 1825–1875* (New York: Oxford University Press, 1995), p. 38.

10 Norman Maclean, *A River Runs Through It, and Other Stories* (Chicago and London: University of Chicago Press, 1976), p. 36.

11 Ibid., pp. 16–17.

12 Ibid., p. 60.

13 Richard Kerridge, 'Small Rooms and the Ecosystem: Environmentalism and DeLillo's *White Noise*', in *Writing the Environment: Ecocriticism and Literature*, ed. Richard Kerridge and Neil Sammells (London and New York: Zed Book Ltd, 1998), p. 191.

14 See J. Gerard Dollar, 'Misogyny in the American Eden: Abbey, Cather, and Maclean', in *Reading the Earth: New Directions in the Study of Literature and the Environment*, ed. Michael P. Branch, Rochelle Johnson, Daniel Patterson and Scott Slovic (Moscow ID: University of Idaho Press, 1998).

15 Robert Redford, quoted in 'Production Notes', *A River Runs Through It* (Columbia Pictures 1992), p. 5.

16 Clips, 'Redford's "River"', *Hollywood Reporter*, 21 September 1992, p. 3.

17 *People*, 26 October 1991, p. 141.

18 'Entrepreneurs Get Their Hooks into Fly Fishing Movie', *Wall Street Journal*, 25 November 1992, p. B2.

19 Gregg Mitman, *Reel Nature: America's Romance with Wildlife on Film* (Cambridge Mass. and London: Harvard University Press, 1999), pp. 130–1.

20 Yi-Fu Tuan, *Passing Strange and Wonderful: Aesthetics, Nature, and Culture* (Washington DC and Covelo: Island Press/Shearwater Books, 1993), pp. 155–6. Godfrey Reggio's *Koyaanisqatsi* (1983) combines in a non-narrative mode the cult of pristine, 'empty' natural landscapes, and their representation as a spectacle of different speeds, as well as the cult of the ecological Indian living, unlike urban dwellers, in 'balance' with nature. See Michael Dempsey, 'Quatsi Means Life: The Films of Godfrey Reggio', *Film Quarterly*, Spring 1989, pp. 2–12.

21 David Nye, *Narratives and Spaces: Technology and the Construction of American Culture* (Exeter: University of Exeter Press, 1997), p. 22.

22 Susan Sontag, *On Photography* (Harmondsworth: Penguin, 1977), p. 111.

23 Neil Evernden, *The Natural Alien* (Toronto, Buffalo and London: University of Toronto Press, 1993), pp. 94–7.

[24] See Jean-Louis Comolli, 'Technique and Ideology: Camera, Perspective, Depth of Field', in *Movies and Methods Vol. II*, and Jean-Louis Baudry 'Ideo-logical Effects of The Basic Cinematographic Apparatus', ed. Bill Nichols (Berkeley and Los Angeles: University of California Press, 1985); Martin Jay, *Downcast Eyes: The Denigration of Vision in Twentieth-Century French Thought* (Berkeley and London: University of California Press, 1994), p. 470.

[25] Ross, *The Chicago Gangster Theory of Life*, p. 122.

[26] Jhan Hochman, *Green Cultural Studies: Nature in Film, Novel, and Theory* (Moscow ID: University of Idaho Press, 1998), p. 3.

[27] Philip Rosen, *Narrative, Apparatus, Ideology: A Film Theory Reader* (New York: Columbia University Press, 1986), p. 282.

[28] Albert Boime, *The Magisterial Gaze: Manifest Destiny and the American Landscape Painting, c. 1830–1865* (Washington DC: Smithsonian Institution, 1991).

[29] Denis Cosgrove, *Social Formation and Symbolic Landscape* (London and Sydney: Croom Helm, 1984), p. 176. See also John Berger, *Ways of Seeing* (London: British Broadcasting Corporation, 1972).

[30] Stuart C. Aitken and Leo E. Zonn, eds, *Place, Power, Situation, and Spectacle: A Geography of Film* (Lanham MD: Rowman and Littlefield Publishers, Inc., 1994), p. 17.

[31] Hochman, *Green Cultural Studies*, p. 6.

[32] Karla Armbruster, 'Creating the World We Must Save: The Paradox of Television Nature Documentaries', in *Writing the Environment: Ecocriticism and Literature*, ed. Kerridge and Sammells, p. 221.

[33] Jane Tompkins, *West of Everything: The Inner Life of Westerns* (New York and Oxford: Oxford University Press, 1992), pp. 74–5.

[34] D.N. Rodowick, *The Crisis of Political Modernism: Criticism and Ideology in Contemporary Film Theory* (Urbana and Chicago: University of Illinois Press, 1999), p. 34.

[35] Tompkins, *West of Everything*, p. 76.

[36] Christine L. Oravec, 'To Stand Outside Oneself: The Sublime in the Discourse of Natural Scenery', in *The Symbolic Earth: Discourse and Our Creation of the Environment*, ed. James G. Cantrill and Christine L. Oravec (Lexington: University Press of Kentucky, 1996), p. 65.

[37] Martin W. Lewis, 'Radical Environmental Philosophy and the Assault on Reason', in *The Flight From Science and Reason*, ed. Paul R. Gross, Norman Levitt and Martin W. Lewis (New York: New York Academy of Sciences, 1996), p. 215.

[38] Kate Soper, *What is Nature? Culture, Politics and the Non-Human* (Oxford and Cambridge Mass.: Blackwell, 1995), pp. 174–5.

Chapter Three

[1] Annette Kolodny, *The Land Before Her: Fantasy and Experience of the American Frontiers, 1630–1860* (Chapel Hill and London: University of North Carolina Press, 1984), p. 8.

[2] Kate Soper, *What is Nature? Culture, Politics and the Non-Human* (Oxford and

Cambridge Mass.: Blackwell, 1995), p. 107.

3 Linda Ruth Williams, 'Blood Brothers', *Sight and Sound*, September 1994, p. 17.

4 J.W. Williamson, *Hillbillyland: What the Movies Did to the Mountains and What the Mountains Did to the Movies* (Chapel Hill and London: University of North Carolina Press, 1995), p. 157.

5 Williams, 'Blood Brothers', p. 17. There are also good accounts of *Deliverance* in Jhan Hochman, *Green Cultural Studies: Nature in Film, Novel, and Theory* (Moscow ID: University of Idaho Press, 1998), pp. 71ff. and Carol Clover, *Men, Women and Chainsaws: Gender in the Modern Horror Film* (London: BFI Publishing, 1992).

6 Fred Pfeil, *White Guys: Studies in Postmodern Domination and Difference* (London: Verso, 1995), p. 49.

7 Gilles Deleuze and Félix Guattari, *A Thousand Plateaus: Capitalism and Schizophrenia*, trans. Brian Massumi (London: Athlone Press, 1988), p. 240.

8 Yvonne Tasker, *Spectacular Bodies: Gender, Genre and the Action Cinema* (London and New York: Routledge, 1993), pp. 132ff.

9 On the varieties of ecofeminism, see Carolyn Merchant, *Radical Ecology: The Search for a Livable World* (New York and London: Routledge, 1992), pp. 183–210. For socialist ecofeminism, see Ariel Salleh, *Ecofeminism as Politics: Nature, Marx and the Postmodern* (London and New York: Zed Books, 1997); on spiritual ecofeminism, see Starhawk, *The Spiral Dance: A Rebirth of the Ancient Religion of the Great Goddess* (San Francisco and London: Harper and Row, 1979).

10 Erik Davis, review of *FernGully*, *Village Voice*, 22 April 1992, p. 63.

11 David Seligman, 'Talking Back to Trees', *Fortune*, 29 June 1992, p. 2. Seligman added that, 'Nexis regrettably failed to confirm our intuition that the mines are centred in the Australian rain forest'.
FAI Insurances Ltd directed part of the profits from the movie to mainstream environmental projects, including the Sierra Club, Greenpeace and the Rain Forest Foundation. This environmental advocacy was accomp-anied by merchandising deals with Pizza Hut, Green Giant and Dial Soap. *FernGully* was also the first feature film ever to be screened in the General Assembly of the United Nations, to commemorate Earth Day on 22 April 1992, thereby gaining for the movie a central place in the iconography of global, mainstream environmentalism.

12 Janet Biehl, *Rethinking Ecofeminist Politics* (Boston: South End Press, 1991), p. 101.

13 Ibid., p. 91.

14 Soper, *What is Nature?*, p. 277.

Chapter Four

1 Neil Evernden, *The Social Creation of Nature* (Baltimore and London: Johns Hopkins University, 1992), p. 25.

2 Martin W. Lewis, *Green Delusions: An Environmentalist Critique of Radical Environmentalism* (Durham and London: Duke University Press, 1992), p. 43.

[3] Ibid., p. 63.

[4] Ibid., pp. 46ff. See, for example, Kirkpatrick Sale, *Dwellers in the Land: The Bioregional Vision* (San Francisco: Sierra Club Books, 1985); Thomas Berry, *The Dream of the Earth* (San Francisco: Sierra Club Books, 1998).

[5] Robert Frank Leslie, *The Bears and I: Raising Three Cubs in the North Woods* (New York: Ballantine Books, 1968).

[6] On the historical context of federal policies towards Indians and National Parks, see Mark David Spence, *Dispossessing the Wilderness: Indian Removal and the Making of National Parks* (New York and Oxford: Oxford University Press, 1999), pp. 133–9.

[7] 'Production Notes', *On Deadly Ground* (Warner Bros., 1994), pp. 4–5.

[8] Anthony Newman, 'Steven Seagal's Box-Office Smash Scorned in Alaska', *Los Angeles Times*, 4 March 1996, p. F6.

[9] 'Production Notes', *On Deadly Ground*, p. 5.

[10] Lewis, *Green Delusions*, p.54.

[11] O. Douglas Schwarz, 'Indian Rights and Environmental Ethics', *Environmental Ethics*, 9.4 (Winter 1989), p. 298.

[12] See Barry Holstun Lopez, *Of Wolves And Men* (New York: Touchstone, 1978), pp. 140ff.

[13] Simon Schama, 'The Princess of Eco-Kitsch', *New York Times*, 14 June 1995, p. A21.

[14] Paula Gunn Allen, *The Sacred Hoop: Recovering the Feminine in American Indian Traditions* (Boston: Beacon Press, 1986).

[15] Robert Tilton, *Pocahontas: The Evolution of an American Narrative* (Cambridge: Cambridge University Press, 1994), p. 63.

[16] *No Sweat: Fashion, Free Trade and the Rights of Garment Workers*, ed. Andrew Ross (New York and London: Verso, 1997), p. 9. See also Pauline Turner Strong, 'Playing Indian in the 1990s', in *Hollywood's Indian: The Portrayal of the Native American in Film*, ed. Peter C. Rollins and John E. O'Connor (Lexington: University Press of Kentucky, 1998), pp. 193–205.

[17] On Leonard Peltier, see John M. Peterson, *Aim on Target: The FBI's War on Leonard Peltier and the American Indian Movement* (John M. Peterson, 1994); Peter Matthiessen, *In the Spirit of Crazy Horse* (New York: Collins-Harvill, 1992); George Sullivan, *Not Guilty: Five Times When Justice Failed* (New York and London: Scholastic Inc., 1997).

[18] For perceptive accounts of the film, see Robert Burgoyne, *Film Nation: Hollywood Looks at U.S. History* (Minneapolis and London: University of Minnesota Press, 1997), pp. 38–56; Terry Wilson, 'Celluloid Sovereignty', in *Legal Reelism: Movies as Legal Texts*, ed. John Denvir (Urbana and Chicago: University of Illinois Press, 1996), pp. 216–23.

Chapter Five

[1] Andrew Ross, *The Chicago Gangster Theory of Life* (London and New York: Verso, 1994), p. 9.

[2] Susanna Hecht and Alexander Cockburn, *The Fate of the Forest* (London: Verso

1989), p. 195. On Hollywood advocacy for the rain forests, see Jane Lieberman, 'Celebs Turn Out in Force For Rainforest Fundraiser', *Variety*, 14 February 1990, p. 14.

3 Andrew Revkin, *The Burning Season* (London: Collins, 1990). See also Nelson Hoineff: 'Puttnam, Rodrigues Huddle on Mendes pic', *Variety*, 11 July 1990, p. 25; David Puttnam, talking on *The Late Show*, BBC 2 television, 1992. *The Burning Season* was eventually filmed as a television movie in 1994.

4 Candace Slater, 'Amazonia as Edenic Narrative', in *Uncommon Ground: Rethinking the Human Place in Nature*, ed. William Cronon (New York and London: W.W. Norton & Company, 1996), p. 117.

5 Ibid., pp. 126, 127.

6 Erik Davis, 'The Jungle', *Village Voice*, 2 June 1992, p. 8.

7 Hecht and Cockburn, *The Fate of the Forest*, p. 11.

8 Ella Shohat and Robert Stam, *Unthinking Eurocentrism* (London: Routledge, 1994), p. 34.

9 Davis, 'The Jungle', p.10.

10 See the comments of Shohat and Stam on both Coppola and Herzog's colonialist attitudes and practices, *Unthinking Eurocentrism*, p. 188.

11 Jack Mathews, 'Into the Rain Forest', *Los Angeles Times*, 18 November 1990, p. 25. However, chief Tariano Ismael Moreiro accused Babenco of providing inadequate medical assistance to tribespeople exposed to diseases such as pneumonia, malaria and fevers. Producer Francisco Ramalho Jr. denied the accusation. See Denis Wright, 'Indian chief slams medical provision on Babenco film', *Screen International*, March 15–21 1991, p. 7.

12 James Brooke, 'In the Brazilian Jungle They Have a Story to Tell', *New York Times*, 19 August 1990, p. 20. See also Neil Okrent, '*At Play in the Fields of the Lord*: An Interview with Hector Babenco', *Cineaste*, 19.1 (1992), pp. 44–7; Elisa Leonelli, 'Hector Babenco', *Venice Magazine*, December 1991, pp. 34–5.

13 Brooke, 'In the Brazilian Jungle They Have a Story to Tell', p. 20.

14 Hecht and Cockburn, *The Fate of the Forest*, p. 11.

15 President Bush refused to sign the biodiversity treaty at the Rio summit in 1992, as it would have bound American companies to pay royalties to the South for genes patented in medical research. The issue of property rights on species, and the commercial exploitation of species, is a contentious one. When genes from the South are owned by transnational corporations, the situation is more one of 'biopiracy' than the utopian, equitable alliance reached at the end of *Medicine Man*. See Tom Athanasiou, *Slow Reckoning: The Ecology of a Divided Planet* (London: Secker and Warburg, 1996), p. 205–11.

16 Hecht and Cockburn, *The Fate of the Forest*, pp. 41, 100.

17 Ibid., pp. 231–2, 98, 122.

18 Athanasiou, *Slow Reckoning*, p. 54.

19 Hecht and Cockburn, *The Fate of the Forest*, p. 196. Andrew Ross points out that this policy does have its native supporters, in *The Chicago Gangster Theory of Life*, p. 9.

20 Ross, *The Chicago Gangster Theory of Life*, p. 9.

21 Slater, 'Amazonia as Edenic Narrative', p. 127.

Introduction to Part Two

1. Roderick Nash, 'The Exporting and Importing of Nature: Nature Appreciation as a Commodity 1850–1980', *Perspectives in American History*, XII (1979), p. 521.
2. David Peterson del Mar, '"Our Animal Friends": Depictions of Animals in *Reader's Digest* During the 1950s', *Environmental History*, 3.1 (January 1998), p. 25.
3. Robert C. Bannister, *Social Darwinism: Science and Myth in Anglo-American Social Thought* (Philadelphia: Temple University Press, 1979), p. 199.
4. Ibid., p. 136.
5. Ralph H. Lutts, *The Nature Fakers: Wildlife, Science and Sentiment* (Golden, CO: Fulcrum Publishing, 1990), pp. 202–3.
6. Robert Elman, *First in the Field: America's Pioneering Naturalists* (London: Van Nostrand Reinhold Co., 1977), p. 209.
7. Ted Benton, *Natural Relations: Ecology, Animal Rights and Social Justice* (London: Verso, 1993), p. 66.
8. Ralph H. Lutts, 'The Wild Animal Story: Animals and Ideas', in *The Wild Animal Story* ed. Ralph H. Lutts (Philadelphia: Temple University Press, 1998), p. 15.
9. For a definition of 'theriophobia', see Barry Holstun Lopez, *Of Wolves and Men* (New York: Touchstone, 1978), p. 140.
10. Bob Mullan and Garry Marvin, *Zoo Culture* (London: Weidenfeld and Nicolson, 1987), p. 7.
11. Michael Ryan and Douglas Kellner, *Camera Politica: The Politics and Ideology of Contemporary Hollywood Film* (Bloomington and Indianapolis: Indiana University Press, 1990), p. 60.
12. Stephen Heath, '*Jaws*, Ideology and Film Theory', *Framework*, 4 (Autumn 1976), pp. 25–7.

Chapter Six

1. Richard Slotkin, *Gunfighter Nation: The Myth of the Frontier in Twentieth-Century America* (New York: Atheneum, 1992), p. 57.
2. See Pete Weisser, 'Hunting in the Movies', *Outdoor California*, November/December 1989, pp. 1–7.
3. Kirkpatrick Sale, *Dwellers in the Land: The Bioregional Vision* (San Francisco: Sierra Club Books, 1985), p. 6.
4. Robert Jewett and John Shelton Lawrence, *The American Monomyth* (Lanham MD: University Press of America, 1988), p. 88.
5. For a comprehensive list of films about Buffalo Bill, see *The BFI Companion to the Western*, ed. Edward Buscombe (London: André Deutsch, 1988), pp. 91–3.
6. Slotkin, *Gunfighter Nation*, p. 14.
7. Jewett and Lawrence, *The American Monomyth*, p.93.
8. Frederick Jackson Turner, *The Frontier in American History* (New York: Henry Holt, 1920).

[9] On the psychology of hunting, see Lewis Regenstein, *The Politics of Extinction: The Shocking Story of the World's Endangered Wildlife* (New York and London: Macmillan, 1975), pp. 39–43.

[10] Margo Kasdan, and Susan Tavernetti, 'The Hollywood Indian in *Little Big Man*: A Revisionist View', *Film and History* XXIII, 1.4 (1993), pp. 75–6.

[11] Martin Padget, 'Film, Ethnography and the Scene of History: *Dances With Wolves* and Participant Observation', *Borderlines* 3.4 (1996), pp. 396–412. Robert Baird carefully distinguishes between the original theatrical release of the film, and the director's 'special edition', in '"Going Indian"', in *Hollywood's Indian*: *The Portrayal of the Native American in Film*, ed. Peter C Rollins and John E. O'Connor (Lexington: University Press of Kentucky, 1998), pp. 153–69.

[12] Dan Flores, 'Bison Ecology and Bison Diplomacy: The Southern Plains from 1800 to 1850', *Journal of American History*, 78.2 (September 1991), pp. 483–5. See also Fergus M. Bordewich, *Killing the White Man's Indian: Reinventing Native Americans at the End of the Twentieth Century* (New York and London: Doubleday, 1996), pp. 208–12.

[13] Shepard Krech III, *The Ecological Indian: Myth and History* (New York and London: W.W. Norton & Company, 1999), p. 149.

[14] Budd Schulberg, *Across The Everglades: A Play for the Screen* (New York: Random House, 1958), 'Introduction', pp. i–xxvi. The movie has become a test case for 'auteur' theory, regarding who had the greater artistic influence, writer Schulberg or director Ray. On Ray's contribution the film, see Bernard Eisenschitz, *Nicholas Ray: An American Journey* (London and Boston: Faber and Faber, 1993), pp. 316–38. See also Geoff Andrew, *The Films of Nicholas Ray: The Poet at Nightfall* (London: Charles Letts, 1991).

[15] Thomas R. Dunlap, *Saving America's Wildlife* (New Jersey: Princeton University Press, 1988), pp. 13ff.

[16] Budd Schulberg, with photographs by Geraldine Brooks, *Swan Watch* (London: Robson Books, 1975), p. 8. For the story of Guy Bradley, 'A Martyr to Millinery', see Charles M. Brookfield and Oliver Griswold, *They All Called It Tropical: True Tales of the Romantic Everglades, Cape Sable, and the Florida Keys* (Miami: Historical Association of Southern Florida, 1949), pp. 60–77.

[17] Schulberg, *Across the Everglades*, pp. xxv–xxvi.

[18] 'Film Opens In Blaze of Glory', *Miami Herald*, 21 August 1958, p. C8. Advertisement, ibid., p. F6.

[19] Herman Melville, *Moby-Dick, or, The Whale* (1850) (Harmondsworth: Penguin, 1972), pp. 637ff.

[20] See Jacques-Yves Cousteau and the staff of the Cousteau Society, *The Cousteau Almanac: An Inventory of Life on Our Water Planet* (New York: Doubleday, 1981), pp. 291–5.

Chapter Seven

[1] See, for example, the documentary 'Cuddly Sharks', *Survival* (Anglia Productions, 1996).

[2] Lee Drummond, *American Dreamtime: A Cultural Analysis of Popular Movies, and*

their Implications for a Science of Humanity (London: Littlefield Adams, 1996), p. 218.

3 Robert Jewett and John Shelton Lawrence, *The American Monomyth* (Lanham MD: University Press of America, 1988), p. 157.

4 Michael Ryan and Douglas Kellner, *Camera Politica: The Politics and Ideology of Contemporary Hollywood Film* (Bloomington and Indianapolis: Indiana University Press, 1990). p. 57.

5 Barbara Creed, *The Monstrous Feminine: Film, Feminism, Psychoanalysis* (London and New York: Routledge, 1993), pp. 27, 107.

6 John Patterson, 'Sequel Opportunities', *Neon*, August 1998, p. 24.

7 Gregg Mitman, *Reel Nature: America's Romance with Wildlife on Film* (Cambridge Mass. and London: Harvard University Press, 1999), p. 177. On Sea World in San Diego, opened in 1964, see Susan G. Davis, *Spectacular Nature: Corporate Culture and the Sea World Experience* (Berkeley and Los Angeles: University of California Press, 1997). On whale-watching as a tourist activity, see Roger Payne, *Among Whales* (New York: Scribner, 1995), pp. 241–2.

8 On dolphins in Greek legend, see Lewis Regenstein, *The Politics of Extinction: The Shocking Story of the World's Endangered Wildlife* (New York and London: Macmillan, 1975), p. 54.

9 See *Zoology: On (Post) Modern Animals*, ed. Bart Verschaffel and Mark Verminck Dublin: The Lilliput Press; Leuven/Amsterdam: Kritak, 1993).

10 'Jacques Yves Cousteau at 85', interview by Jim Motavelli and Susan Elan, *E Magazine*, March–April 1996, p. 11.

11 Bill McKibben, *The Age of Missing Information* (New York: Random House, 1992), p. 73. For the Marine Mammal Protection Act, see Regenstein, *The Politics of Extinction*, pp. 100–17; Victor B. Scheffer, *The Shaping of Environ-mentalism in America* (Seattle and London: University of Washington Press, 1991), p. 162; Samuel P. Hays, *Beauty, Health, and Permanence: Environ-mental Politics in the United States, 1955–1985* (Cambridge, New York etc.: Cambridge University Press, 1987), p. 113.

12 In *The Day of the Dolphin* (1973), dolphins are taught to speak by a scientist (George C. Scott), who later comes to regret his exploitation of them, when they are caught up as bomb carriers in a terrorist plot to assassinate the President.

13 Ivan Tors, *My Life in the Wild* (Boston: Houghton Mifflin Co., 1979), pp. 188, 187. See also, 'Gentle Movie "Killer" Dies', *Los Angeles Herald-Examiner*, 7 October 1966; Robert Lansing, 'Namu: Nice Guy Killer Whale', *Los Angeles Times*, Calendar, 24 July 1966.

14 Bob Mullan and Garry Marvin, *Zoo Culture* (London: Weidenfeld and Nicolson, 1987), pp. 21, 23.

15 Michel Foucault, *Discipline and Punish: The Birth of the Prison*, trans. Alan Sheridan (London: Allen Lane, 1977), p. 136.

16 Tom Regan, *All That Dwells Therein: Animal Rights and Environmental Ethics* (Berkeley: University of California Press, 1982). See also Ted Benton, *Natural Relations: Ecology, Animal Rights and Social Justice* (London: Verso, 1993), p. 3.

17 On the Free Keiko campaign, see Mitman, *Reel Nature*, p. 158.

[18] '*Free Willy* Producers Join Blockade', *Los Angeles Times*, 26 November 1993, p. F2.

[19] Gini Kopecky, '"Free Willy" . . . and Maybe Rescue Keiko, Too', *New York Times*, 11 July 1993, p. 10.

Chapter Eight

[1] Lewis Regenstein, *The Politics of Extinction*: *The Shocking Story of the World's Endangered Wildlife* (New York and London: Macmillan, 1995), p. 183.

[2] William J. Long, *Mother Nature: A Study of Animal Life and Death* (New York and London: Harper and Bros, 1923), p. 25. See also Ralph H. Lutts, *The Nature Fakers: Wildlife, Science and Sentiment* (Golden CO: Fulcrum Publishing, 1990), p. 92–3; Lisa Mighetto, *Wild Animals and Environmental Ethics* (Tucson: University of Arizona Press, 1991), pp. 89–90.

[3] Barry Holstun Lopez comments that 'whatever ill they might have believed of wolves, most people also believed they were devoted and affectionate parents'. *Of Wolves and Men* (New York: Touchstone, 1978), p. 244.

[4] Richard Schickel, *The Disney Version* (London: Pavilion, 1986), pp. 153–5.

[5] Ernest Thompson Seton, *Wild Animals I Have Known* (1898) (Harmondsworth: Penguin, 1987).

[6] Rick McIntyre, 'Introduction', *War Against the Wolf: America's Campaign to Exterminate the Wolf*, ed. Rick McIntyre (Stillwater MN: Voyager Press, 1995), pp. 76–7. See also Lopez, *Of Wolves and Men*, p. 191.

[7] Lopez, *Of Wolves and Men*, p. 266.

[8] McIntyre, *War Against the Wolf*, p. 217.

[9] Seton, *Wild Animals I Have Known*, p. 51.

[10] Ibid., p. 20.

[11] For the True-Life Adventure series, see Alexander Wilson, *The Culture of Nature*: *North American Landscape from Disney to Exxon Valdez* (Oxford: Blackwell, 1992), pp. 117–21; Gregg Mitman, *Reel Nature*: *America's Romance with Wildlife on Film* (Cambridge Mass. and London: Harvard University Press, 1999), p. 109–31.

[12] Seton, *Wild Animals I Have Known*, pp. 17, 18, 49, 53–4, 12.

[13] James Powers, 'Disney Production Holds Wide Appeal', *Hollywood Reporter*, 24 October 1962, p. 3; Hazel Flynn, 'You'll Laugh, Cry Over Wolf Hero', *Citizen-News*, 10 November 1962 (*Legend of Lobo* cuttings file, Margaret Herrick Library of the Academy of Motion Pictures, Arts and Sciences, Los Angeles); Bosley Crowther, 'The Wolf Becomes a Hero', *New York Times*, 10 November 1962, p. 16.

[14] Farley Mowat, *Never Cry Wolf* (New York: Dell, 1963).

[15] Stephen Hunter, '*Never Cry Wolf* Offers Viewers the Spectacular Wonder of Nature', *Baltimore Sun*, 11 November 1983, p. B4.

[16] John Noble Wilford, 'A New Movie Boasts Unlikely Stars—Wolves', *New York Times*, 9 October 1983, p. 21.

[17] Daniel B. Botkin, *Discordant Harmonies*: *A New Ecology for the Twenty-First Century* (New York and Oxford: Oxford University Press, 1990), p. 47. On the impact

of hunting and 'wildlife management' policies on wolf populations in North America, see Charles Bergman, *Wild Echoes: Encounters with the Most Endangered Animals in North America* (Anchorage and Seattle: Alaska Northwest Books, 1990), pp. 12ff.; Lewis Regenstein, *The Politics of Extinction*, pp. 13ff.

[18] Robert Ornstein and Paul Ehrlich, *New World, New Mind: Moving Toward Conscious Evolution* (New York: Simon and Schuster, 1989), p. 59.

[19] Marykay Powell, in 'Production Notes', *White Fang* (Walt Disney Productions, 1991), p. 18.

[20] Terry Pristin, 'The Filmmakers vs. The Crusaders', *Los Angeles Times*, Calendar, 29 December 1991, p. 30.

[21] Lopez, *Of Wolves and Men*, p. 4. The assertion made by Defenders of Wildlife appears, for example, in Regenstein, *The Politics of Extinction*, p. 183.

[22] 'Film Shorts', *Hollywood Reporter*, 26 June 1991, p. 18.

[23] Jack London, *The Call of the Wild, White Fang, and Other Stories* (Harmondsworth: Penguin, 1981), pp. 206, 233.

[24] 'Production Notes', *White Fang*, p. 8.

[25] Lopez, *Of Wolves and Men*, p. 170.

[26] The sequel *White Fang 2* (1994) explored the relationship between the wolf, the Haida tribe and rapacious American gold prospectors. The final scene reiterates the image of the wolf as a family animal raising new born cubs, in a reassuring image of the family of nature.

[27] For the symbolic significance of the reintroduction of the wolf to Yellowstone, see the article by Secretary of the US Department of the Interior Bruce Babbitt, 'The Fierce Green Fire', *Audubon*, September/October 1994, p. 120; Rodger Schlickeisen, 'Wolf Recovery's Significance', *Half a Hundred: Defenders Magazine*, Winter 1996–7, from Defenders of Wildlife web-site, http://www.defenders.org, 18 September 1997.

[28] Disney's use of music is discussed in Ross B. Care, 'Threads of Melody—The Evolution of a Major Film Score—Walt Disney's *Bambi*', *Quarterly Journal of the Library of Congress*, Spring 1983, pp. 79–98.

[29] Robert Franklin Leslie, *The Bears and I*, pp. 179–80, 181–2.

[30] Regenstein, *The Politics of Extinction*, pp. 194–208. On the history and symbolic importance of bears in North America, see Paul Shepard and Barry Sanders, *The Sacred Paw: The Bear in Nature, Myth, and Literature* (New York: Viking, 1985), pp. 94–101; R. Edward Grumbine, *Ghost Bears: Exploring the Biodiversity Crisis* (Washington DC and Covelo: Island Press, 1992), pp. 64–86.

[31] Regenstein, *The Politics of Extinction*, pp. 195ff.

Chapter Nine

[1] Ella Shohat and Robert Stam, *Unthinking Eurocentrism: Multiculturalism and the Media* (London and New York: Routledge, 1994), p. 104.

[2] Thomas Delehanty, 'The New Film', *Post*, 2 April 1931 (*Trader Horn* cuttings file, Margaret Herrick Library of the Academy of Motion Pictures, Arts and Sciences, Los Angeles).

[3] Shohat and Stam, *Unthinking Eurocentrism*, p. 87. See also Fatimah Tobing Rony,

The Third Eye: Race, Cinema and Ethnographic Spectacle (Durham and London: Duke University Press, 1996).

4 For the gender and racial implications of the film, see Walt Morton, 'Tracking the Sign of Tarzan: Trans-media Representation of a Pop-culture Icon', in *You Tarzan: Masculinity, Movies and Men*, ed. Pat Kirkham and Janet Thumin (London: Lawrence and Wishart Ltd, 1993), pp. 106–25; Rhoda J. Berenstein, *Attack of the Leading Ladies: Gender, Sexuality, and Spectatorship in Classic Horror Cinema* (New York: Columbia University Press, 1996), pp. 160ff.

5 On Hollywood adaptations of Hemingway's safari stories, see Robert E. Morsberger, '"That Hollywood Kind of Love": Macomber in the Movies', and Gene D. Phillips, 'Novelist versus Screenwriter: The Case for Casey Robinson's Adaptations of Hemingway's Fiction', in *A Moving Picture Feast: The Filmgoer's Hemingway*, ed. Charles M. Oliver (New York, Westport and London: Praegar, 1989).
Romain Gary, *The Roots of Heaven* (London: Michael Joseph, 1958). In his autobiography, John Huston makes no comment on how a former big game hunter like himself came to admire Gary's novel. For Huston's comments on the making of *The Roots of Heaven,* see *An Open Book* (New York: Alfred A. Knopf, 1980), pp. 274–80. Nor does his biographer comment on the issue. See Axel Madsen, *John Huston* (London: Robson Books, 1979), pp. 173–9. The made-for-cable movie *The Last Elephant* (1990) also deals with the issue of elephant conservation.

6 Joseph McBride, *Hawks on Hawks* (London: University of California Press, 1982), p. 147; Peter Bogdanovich, 'Interview with Howard Hawks', in *Howard Hawks: American Artist*, ed. Jim Hillier and Peter Wollen (London: BFI Publishing, 1996), p. 66.

7 Jean Douchet, '*Hatari!*', in *Howard Hawks: American Artist*, ed. Hillier and Wollen, p. 81.

8 *The Lion Hunters* (1951) was an anti-hunting fable in the children's Bomba series, with children's fiction, it seems, again at the forefront of changing attitudes to nature. I have been unable to find a print of this film.

9 Joy Adamson, *Born Free: A Lioness of Two Worlds* (London: Fontana/Collins, 1960); Sandy Gall, *George Adamson: Lord of the Lions* (London: Grafton Books, 1991), p. 124.

10 See also Joy Adamson, 'What Animals Can Teach Us', *Family Weekly*, 9 October 1966, p. 4.

11 Carl Foreman, 'Foreman the Lion Tamer Extolls on Bearding the Beast on Film', *Variety Weekly*, 13 January 1965, p. 22.

12 'Born Free', *Look*, 19 April 1966, p. 109.

13 Gregg Mitman interprets Elsa as an allegory of African independence: 'Elsa gained her freedom, but as illustrated in her return with her cubs to the Adamsons' camp a year after her release, not at the expense of friendly contact with whites. Westerners projected the same hopes on Kenya as a whole.' *Reel Nature: America's Romance with Wildlife on Film* (Cambridge Mass. and London: Harvard University Press, 1999), p. 201.

14 Gall, *George Adamson*, pp. 138–9; Adrian House, *The Great Safari: The Lives of*

George and Joy Adamson (London: Harvill, 1993), pp. 277–82.

[15] House, *The Great Safari*, pp. 333–8, 367.

[16] Ibid., p. 370.

[17] The success of *Born Free* led to a television series, and to a movie sequel, *Living Free* (1972), which followed the story of Elsa's cubs. The plot concerns African poachers whose activities are considered at odds with the interests of conservationism. The made-for-television *Born Free: A New Adventure* (1996) remade the Elsa story for a new generation, this time with more extensive contributions from African characters than in the first movie.

[18] See Lee Grant, '"Roar" in Danger of Fading Away', *Variety*, 19 July 1978, p. 13.

[19] George F. Custen, *Bio/Pics: How Hollywood Constructed Public History* (New Brunswick: Rutgers University Press, 1992).

[20] Nina J. Easton, 'Film Makers in the Mist', *Los Angeles Times*, 16 September 1988, p. 28.

[21] Tania Modleski argues that criticism of the film which concentrates on the question of its accuracy as a representation of Fossey's life arises 'as a response to the disturbances created *by* the film at a phantasmatic level, instilling in us a longing for an authentic human life to serve as ground and source of the film's meaning'. To discuss the movie in relation to its historical referent is to disavow what she sees as the movie's perverse interest in repressed bestiality and 'bizarre psychosocial dynamics'. However, by disallowing the exploration of the relationship between the film and its historical referent, in particular Fossey's role in conservationism, Modleski effectively prevents an ecocritical discussion of the movie. See Tania Modleski, 'Cinema and the Dark Continent: Race and Gender in Popular Film', in *Feminism Without Women: Culture and Criticism in a 'Post-feminist' Age* (New York and London: Routledge, 1991), pp. 121–6. For an analysis of the film that does take into account its historical referent, see Donna Haraway, *Primate Visions: Gender, Race, and Nature in the World Modern Science* (London and New York: Verso, 1989), pp. 149, 267–8.

[22] Harold T.P. Hayes, *The Dark Romance of Dian Fossey* (New York: Simon and Schuster, 1990), pp. 295–7.

[23] Dian Fossey, *Gorillas in the Mist* (Harmondsworth: Penguin, 1983), pp. 241–2. Jonathan S. Adams and Thomas O. McShane add that the Project is 'not without problems, particularly the continued prominent role of expatriates, the lack of revenue sharing with local communities, and the incomplete transfer of information and expertise to Rwandans'. *The Myth of Wild Africa: Conservation Without Illusion* (London: University of California Press, 1992), p. 204. On the consequences of the Rwandan civil war for gorilla conservation, see ibid., p. 261, and the television documentary *Mountain Gorilla: A Shattered Kingdom* in the *Survival* series, photographed, produced and directed by Bruce Davidson (Anglia Productions, 1996).

[24] Vera Norwood, *Made From the Earth: American Women and Nature* (Chapel Hill: University of North Carolina Press, 1993), p. 210.

[25] Alan Root, speaking in *The Real Dian Fossey* (Cunliffe-Franklyn Productions, National Geographic in association with Channel Four Television, 1998).

[26] Fossey, *Gorillas in the Mist*, pp. 71–89.

[27] House, *The Great Safari*, p. 346.
[28] Adams and McShane, *The Myth of Wild Africa*, p. xv.

Introduction to Part Three

[1] Raymond Williams, *The Country and the City* (London: Chatto and Windus, 1973), pp. 13–34.
[2] Fredric Jameson, *Postmodernism, or, The Cultural Logic of Late Capitalism* (London and New York: Verso, 1991), p. ix.
[3] Donald Worster, *Under Western Skies: Nature and History in the American West* (New York and Oxford: Oxford University Press, 1992), pp. 250–4.
[4] Kate Soper, 'Nature/nature', in *FutureNatural: Nature, Science, Culture*, ed. George Robertson, Melinda Mash, Lisa Tickner, Jon Bird, Barry Curtis and Tim Putnam (London and New York: Routledge, 1996), p. 30.
[5] Verena Andermatt Conley, in *Ecopolitics: The Environment in Poststructuralist Thought* (London and New York: Routledge, 1997), p. 100.
[6] Jhan Hochman, *Green Cultural Studies: Nature in Film, Novel, and Theory* (Moscow ID: University of Idaho Press, 1998), p. 1.
[7] Greg Easterbrook, *A Moment on the Earth: The Coming of Age of Environmental Optimism* (Harmondsworth: Penguin, 1995), p. 376
[8] Williams, *The Country and the City*, pp. 294–5.

Chapter Ten

[1] Jane Tompkins, *West of Everything*: *The Inner Life of Westerns* (New York and Oxford: Oxford University Press, 1992), pp. 114, 118.
[2] Donald Worster, *Under Western Skies: Nature and History in the American West* (New York and Oxford: Oxford University Press, 1992), pp. 34ff.
[3] Ibid., p. 6; see also Henry Nash Smith, *Virgin Land* (Cambridge Mass.: Harvard University Press, 1950).
[4] Worster, *Under Western Skies*, p. 39ff.
[5] Barbara Klinger, '"The Road to Dystopia"': Landscaping the Nation in *Easy Rider*', in *The Road Movie Book*, ed. Steven Cohan and Ina Rae Hark (London and New York: Routledge, 1997), p. 183.
[6] The scenes depicting the Texas Panhandle were actually shot in Canada. See Nestor Almendros, 'Photographing "Days of Heaven"', *American Cinematographer*, June 1979, p. 562.
[7] On the cinematography in the film, see Richard Patterson, 'The River', *American Cinematographer*, November 1994, pp. 64–72.
[8] See Duncan Webster, *Looka Yonder: The Imaginary America of Populist Culture* (London: Comedia, 1988), pp. 29ff., 67–73.
[9] On the FHA and the 1980s farm crisis, see Willard W. Cochrane, *The Development of American Agriculture: A Historical Analysis* (Minneapolis and London: University of Minnesota Press, 1993), pp. 316–22; Victor Davis Hanson, *Fields Without Dreams: Defending the Agrarian Idea* (New York and London: The Free Press, 1996).

[10] For a further comparison of the two movies, see Terry Christensen, *Reel Politics: American Political Movies from 'Birth of a Nation' to 'Platoon'* (Oxford and New York: Basil Blackwell, 1987), pp. 165–7.

[11] John M. Wilson, 'Widening Film's Culture', *Writer's Digest*, March 1987, p. 48.

[12] John Nichols, 'Bob and the Bean Stalk', *American Film*, May 1987, p. 13.

[13] Jill Kearney, 'The Old Gringo', *American Film*, March 1988, p. 67.

[14] 'Bored With His Name', Robert Redford interviewed by Stephen Schaefer, *Film Comment*, February 1988, p. 36.

[15] For the history of the Owens Valley land grab, see Marc Reisner, *Cadillac Desert: The American West and Its Disappearing Water* (London: Secker and Warburg, 1986), pp. 54–107; Jacoba Atlas, 'The Facts behind "Chinatown"', *Los Angeles Free Press*, 27 September 1974, p. 23; Michael Eaton, *Chinatown* (London: BFI Publishing, 1997), pp. 22–7.

[16] Roman Polanski, 'The Day I Gave Jack Nicholson a Bloody Nose', *Gentleman's Quarterly*, February 1984, p. 180.

[17] Eaton, *Chinatown*, p. 27.

[18] Atlas, 'The facts behind "Chinatown"', p. 23.

[19] Robert Towne, 'It's only LA, Jake', *Los Angeles Times*, 29 May 1994, p. 12.

[20] Eaton, *Chinatown*, p. 65.

[21] On *Chinatown* and the detective genre, see John W. Cawelti, '*Chinatown* and Generic Transformation in Recent American Film', in *Film Genre Reader II*, ed. Barry K. Grant (Austin University of Texas Press, 1995), pp. 227–45.

[22] Colin McArthur, 'Chinese Boxes and Russian Dolls: Tracking the Elusive Cinematic City', in David B. Clarke, *The Cinematic City* (London and New York: Routledge, 1997), p. 26.

Chapter Eleven

[1] Eric Mottram, *Blood on the Nash Ambassador: Investigations in American Culture* (London: Hutchinson Radius, 1983), p. 66.

[2] Richard Laermer, 'Cars Don't Have to Take a Back Seat in the Movies', *New York Times*, 20 November 1988, p. H21.

[3] David Nye, *Narratives and Spaces: Technology and the Construction of American Culture* (Exeter: University of Exeter Press, 1997), pp. 77–87.

[4] Ibid., p. 80.

[5] Anthony Newman, 'Steven Seagal's Box-Office Smash Scorned in Alaska', *Los Angeles Times*, 4 March 1994, p. F6.

[6] Shepard Krech III, *The Ecological Indian: Myth and History* (New York and London: W.W. Norton & Company, 1999), pp. 225–6.

[7] Roderick Nash, *Wilderness and the American Mind* (London: Yale University Press, 1982), pp. 272ff.; Peter Coates, *The Trans-Alaska Pipeline Controversy* (London: Bethlehem Lehigh University Press, 1991), pp. 226, 314–15.

[8] Tom Tunney, review of *On Deadly Ground*, *Sight and Sound*, May 1994, p. 48.

[9] See, for example, Ariel Salleh, *Ecofeminism as Politics: Nature, Marx and the Postmodern* (London and New York: Zed Books, 1997); Brian Easlea, *Fathering*

the Unthinkable: Masculinity, Scientists and the Nuclear Arms Race (London: Pluto Press, 1983).

[10] David Pepper, *Eco-Socialism: From Deep Ecology to Social Justice* (London and New York: Routledge, 1993), p. 34.

[11] Paul Gross and Norman Levitt, *Higher Superstition: The Academic Left and Its Quarrels with Science* (Baltimore: Johns Hopkins University Press, 1994), p. 178.

[12] Donald Worster links the *Exxon Valdez* disaster to the government's relaxation of rules concerning double-hulled tankers, in *Under Western Skies: Nature and History in the American West* (New York and Oxford: Oxford University Press, 1992), p. 223.

[13] See Philip S. Egan, *Design and Destiny: The Making of the Tucker Automobile* (Orange: On the Mark, 1989), pp. 109–18.

[14] See Gerrylynn K. Roberts and Philip Steadman, *American Cities and Technology: Wilderness to Wired City* (London: Routledge, 1999), pp. 76–8.

[15] John Wrathall, review of *The American President*, *Sight and Sound*, January 1996, p. 36.

[16] See Tom Wicker, 'Waiting for an Environmental President', *Audubon*, September/October 1994, pp. 49–103.

Chapter Twelve

[1] See Paul Boyer, *By the Bomb's Early Light: American Thought and Culture at the Dawn of the Atomic Age* (New York: Pantheon Books, 1985).

[2] On nuclear war movies, see Kim Newman, *Millennium Movies: End of the World Cinema* (London: Titan, 1999); Joyce A. Evans, *Celluloid Mushroom Clouds: Hollywood and the Atomic Bomb* (Boulder and Oxford: Westview, 1998); Mike Broderick, *Nuclear Movies: A Critical Analysis and Filmography* (Jefferson NC: McFarland, 1991); Toni Perrine, *Film and the Nuclear Age: Representing Nuclear Anxiety* (New York and London: Garland, 1998); Jack Shaheen, *Nuclear War Films* (Carbondale and London: Southern Illinois University, 1978); Ralph H. Lutts, 'Chemical Fallout and the Environmental Movement', in *The Recent Past: Readings on America Since World War II* ed. Allan M. Winkler (New York: Harper and Row, 1989), pp. 326–38.

[3] Spencer Weart, *Nuclear Fear: A History of Images* (Cambridge Mass. and London: Harvard University Press, 1988), p. 350.

[4] Joseph G. Morone and Edward J. Woodhouse, *The Demise of Nuclear Energy? Lessons for Democratic Control of Technology* (New Haven and London: Yale University Press, 1989), p. 86.

[5] Michael Ryan and Douglas Kellner, *Camera Politica: The Politics and Ideology of Contemporary Hollywood Film* (Bloomington and Indianapolis: Indiana University Press, 1990), p. 104.

[6] Gerald H. Clarfield and William M. Wiecek, *Nuclear America: Military and Civil Nuclear Power in the United States 1940–1980* (New York: Harper and Row, 1984), pp. 388–9. On the linking of *The China Syndrome* and Three Mile Island, see Aaron Latham, 'Hollywood vs. Harrisburg', *Esquire*, 22 May 1979, pp. 77–86.

[7] David Burnham 'Nuclear Experts Debate "The China Syndrome"', *New York Times*, 18 January 1979, p. 1.

[8] Ibid., p. 19.

[9] Robert Gillette, '"China Syndrome": Nuclear Reactions', *Los Angeles Times*, 25 March 1979, p. 3.

[10] Burnham, 'Nuclear Experts Debate "The China Syndrome"', p. 19.

[11] Clarfield and Wiecek, *Nuclear America*, pp. 377–8.

[12] Ibid., p. 349.

[13] Gillette, '"China Syndrome": Nuclear Reactions', p. 1.

[14] Weart, *Nuclear Fear*, p. 353.

[15] For a perceptive analysis of both *The China Syndrome* and *Silkwood*, see Terry Christensen, *Reel Politics: American Political Movies from 'Birth of a Nation' to 'Platoon'* (Oxford and New York: Basil Blackwell, 1987), pp. 161–64.

[16] Hayden White, 'The Modernist Event', in *The Persistence of History: Cinema, Television and the Modern Event* (New York: Routledge, 1996), pp. 20, 21, 23. For a critique of postmodern attitudes to the nuclear age, see J. Fisher Solomon, *Discourse and Reference in the Nuclear Age* (Norman: University of Oklahoma Press, 1988).

[17] Linda Williams, 'Melodrama Revisited', in *Refiguring American Film Genres: Theory and History*, ed. Nick Browne (Berkely, Los Angeles and London: University of California Press, 1998), p. 42.

Conclusion

[1] Michael Wood, *America in the Movies, or, "Santa Maria, It Had Slipped My Mind"* (New York: Columbia Press, 1975), p. 15.

[2] Richard Maltby and Ian Craven, *Hollywood Cinema* (Oxford and Cambridge Mass.: Blackwell, 1995), p. 457.

[3] On issues of racial and ethnic identity in the film, see Angela Aleiss, 'Le Bon Sauvage: *Dances With Wolves* and the Romantic Tradition', *American Indian Culture and Research Journal*, 15.4 (1991), pp. 91–6; Randall A. Lake, 'Argument-ation and Self: The Enactment of Identity in *Dances With Wolves*', *Argumentation and Advocacy*, 34 (Fall 1997), pp. 66–89; Robert Baird, '"Going Indian"', in *Hollywood's Indian: The Portrayal of the Native American in film*, ed. Peter C. Rollins and John E. O'Connor (Lexington: University Press of Kentucky, 1998), pp. 153–69. On the relationship between the film and a history of the Plains Indians, see Wayne Michael Sarf, 'Oscar Eaten by Wolves', *Film Comment*, November/December 1991, pp. 62–70; Fergus M. Bordewich, *Killing the White Man's Indian: Reinventing Native Americans at the End of the Twentieth Century* (New York and London: Doubleday, 1996), pp. 13, 28–30; Meredith Berkman, 'How "Dances" Got Real', *Entertainment*, 8 March 1991, p. 22. On the Sioux as a contemporary Indian people, see George Miller, 'Progress for Indians is a Film Fantasy', *Los Angeles Times*, 26 March 1991, p. B7; Judith Valente, 'A Century Later, Sioux Still Struggle, And Still Are Losing', *Wall Street Journal*, 25 March 1991, pp. A1–2.

Steven M. Leuthold observes that some Native American audiences praised the film, suggesting that they thereby viewed 'the idealization of native culture as a positive aspect of representation, because it serves as a counter-weight to negative stereotypes'. 'Native American Responses to the Western', *American Indian Culture and Research Journal*, 19.1 (1995), p. 181. For positive Native American reactions to the film, see Tim Giago, Mary Cook and Gemma Lockhart, 'They've Gotten It Right This Time', *Native Peoples*, Winter 1991, pp. 6–14.

4 Henri Lefebvre, *The Production of Space*, trans. Donald Nicholson-Smith (Oxford and Cambridge Mass.: Blackwell, 1991), p. 31.

Filmography

Source: Internet Movie Data Base, www.imdb.com

20,000 Leagues Under the Sea. Dir. Richard Fleischer. Walt Disney Productions. 1954.

The Adventures of Robin Hood. Dirs. Michael Curtiz, William Keighley. First National/ Warner Bros. 1938.

The African Queen. Dir. John Huston. Horizon/Romulus. 1951.

Aguirre: The Wrath of God. Dir. Werner Herzog. Werner Herzog Filmproduktion. 1972.

Alaska. Dir. Fraser C. Heston. Castle Rock/Columbia /Fuchs-Burg. 1996.

Aliens. Dir. James Cameron. 20th Century Fox/Brandywine/Concorde. 1986.

American Graffiti. Dir. George Lucas. Lucasfilm/The Coppola Co./Universal. 1973.

The American President. Dir. Rob Reiner. Castle Rock/Universal/Columbia/ Wildwood. 1995.

Andre. Dir. George Miller. Kushner-Locke/Paramount. 1994.

Apocalypse Now. Dir. Francis Ford Coppola. Zoetrope. 1979.

At Play in the Fields of the Lord. Dir. Hector Babenco. Saul Zaentz. 1991.

Bambi. Dir. David Hand. Walt Disney Productions. 1942.

The Bear (*L'Ours*). Dir. Jean-Jacques Annaud. Renn/Price. 1988.

The Bears and I. Dir. Richard McEveety. Walt Disney Productions. 1974.

The Big Trees. Dir. Felix E. Feist. Warner Bros. 1952.

Bill and Ted's Excellent Adventure. Dir. Stephen Herek. Interscope/Nelson. 1989.

Bio-Dome. Dir. Jason Bloom. Motion Picture Corporation of America. 1996.

Blade Runner. Dir. Ridley Scott. Blade Runner Partnership/The Ladd Company. 1982.

Born Free. Dir. James Hill. Columbia/Open Road/High Road/Atlas 1966.

Born Free: A New Adventure (TV). Dir. Tommy Lee Wallace. Columbia/Tristar. 1996.

Buffalo Bill. Dir. William A. Wellman. 20th Century Fox. 1944.

Buffalo Bill and Escort. Edison Manufacturing Company. 1897.

The Burning Season (TV). Dir. John Frankenheimer. Home Box Office. 1994.

Bwana Devil. Dir. Arch Oboler. Oboler/United Artists. 1952.

The Call of the Wild. Dir. William A. Wellman. 20th Century. 1935.

Call of the Wild. Dir. Ken Annakin. Massfilms/CCC/Izaro/Oceania/UPF/MGM/ Towers of London/Norsk. 1972.

The China Syndrome. Dir. James Bridges. IPC/Columbia. 1978.
Chinatown. Dir. Roman Polanski. Paramount/Long Road/Penthouse. 1974.
Choke Canyon. Dir. Chuck Bail. Brouwersgracht. 1986.
A Civil Action. Dir. Steven Zaillian. Paramount/Touchstone/Wildwood. 1998.
Clarence, the Cross-Eyed Lion. Dir. Andrew Marton. Ivan Tors/MGM. 1965.
Claws. Dir. Robert E. Pierson. Alaska. 1977.
Congo. Dir. Frank Marshall. Kennedy-Marshall/Paramount. 1995.
Country. Dir. Richard Pearce. Walt Disney Productions. 1984.
Crimson Tide. Dir. Tony Scott. Hollywood/Don Simpson-Jerry Bruckheimer. 1995.
Dances With Wolves. Dir. Kevin Costner. Tig/Majestic. 1990.
Day of the Animals. Dir. William Girdler. Film Ventures International. 1977.
Days of Heaven. Dir. Terrence Malick. Paramount. 1978.
The Deer Hunter. Dir. Michael Cimino. Universal/EMI. 1978.
Deliverance. Dir. John Boorman. Warner Bros. 1972.
Easy Rider. Dir. Dennis Hopper. Pando/Columbia/BBS/Raybert. 1969.
The Emerald Forest. Dir. John Boorman. Embassy. 1985.
Fail-Safe. Dir. Sidney Lumet. Columbia. 1964.
Falling Down. Dir. Joel Schumacher. Alcor/LE Studio Canal +/Regency/Warner Bros. 1993.
FernGully: The Last Rainforest. Dir. Bill Kroyer. FAI. 1992.
Fitzcarraldo. Dir. Werner Herzog. Herzog/Project Film/Wildlife Films/ZDF. 1982.
Flipper. Dir. James Clark. MGM. 1963.
Flipper. Dir. Alan Shapiro. Universal Pictures/The Bubble Factory/American Films. 1996.
Flipper's New Adventure. Dir. Leon Benson. Ivan Tors/MGM. 1964.
Fly Away Home. Dir. Carroll Ballard. Columbia/Sandollar. 1996.
The Forbidden Dance. Dir. Greydon Clark. 21st Century Film/Sawmill. 1990.
The Formula. Dir. J.G. Avildsen. MGM/CIP. 1980.
Free Willy. Dir. Simon Wincer. Warner Bros./Alcor/Le Studio Canal +/Regency. 1993.
Free Willy 2: The Adventure Home. Dir. Dwight Little. Warner Bros. 1995.
Free Willy 3: The Rescue. Dir. Sam Pillsbury. Warner Bros./Regency/Shuler-Donner-Donner. 1997.
The Ghost and the Darkness. Dir. Stephen Hopkins. Paramount/Constellation/Douglas-Reuther. 1996.
Godzilla. Dir. Roland Emmerich. TriStar/Centropolis/Fried/Independent. 1998.
Gorillas in the Mist. Dir. Michael Apted. Guber-Peters. 1988.
The Grapes of Wrath. Dir. John Ford. 20th Century Fox. 1940.
Greystoke: The Legend of Tarzan, Lord of the Apes. Dir. Hugh Hudson. Edgar Rice Burroughs Inc./Warner Bros./WEA. 1984.
Grizzly. Dir. William Girdler. Columbia. 1976.
The Grub Stake: A Tale of the Klondike. Dir. Bert Van Tuyle. Nell Shipman Productions. 1923.
Hatari! Dir. Howard Hawks. Malaban. 1962.
Heaven's Gate. Dir. Michael Cimino. United Artists. 1980.
Hiroshima Mon Amour. Dir. Alain Resnais. Argos/Como/Daiei Studios/Pathé. 1959.

How the West Was Won. Dirs. John Ford, Henry Hathaway, George Marshall, Richard Thorpe. Cinerama/MGM. 1963.
Hud. Dir. Martin Ritt. Salem-Dover/Paramount. 1963.
Incident at Oglala. Dir. Michael Apted. North Fork. 1991.
The Iron Horse. Dir. John Ford. Fox Film Corporation. 1924.
It Came From Beneath the Sea. Dir. Robert Gordon. Clover/Columbia. 1955.
Jaws. Dir. Stephen Spielberg. Universal/Zanuck/Brown. 1975.
Jaws 2. Dir. Jeannot Szwarc. Universal. 1978.
Jaws 3D. Dir. Joe Alves. Alan Landsburg/Universal. 1983.
Jaws: The Revenge. Dir. Joseph Sargent. Universal. 1987.
Jeremiah Johnson. Dir. Sydney Pollack. Sanford/Warner Bros. 1972.
King of the Grizzlies. Dir. Ron Kelly. Walt Disney Productions. 1970.
King Solomon's Mines. Dir. Compton Bennett. MGM. 1950.
Koyaanisqatsi. Dir. Godfrey Reggio. Island Alive/Institute for Regional Education. 1983.
The Last Elephant (TV). Dir. Joseph Sargent. Turner Pictures/National Audubon Society. 1990.
The Last Hunt. Dir. Richard Brooks. MGM. 1956.
The Last of the Mohicans. Dir. Michael Mann. Morgan Creek. 1992.
The Legend of Lobo. Dir. James Algar. Walt Disney Productions. 1962.
The Lion King. Dir. Roger Allers. Walt Disney Productions. 1994.
Little Big Man. Dir. Arthur Penn. Cinema Center 100/Stockbridge-Hiller. 1970.
Living Free. Dir. Jack Couffer. Highroad/Open Road. 1972.
The Macomber Affair. Dir. Zoltan Korda. United Artists/Award. 1947.
Mara of the Wilderness. Dir. Frank McDonald. Unicorn. 1966.
Medicine Man. Dir. John McTiernan. Cinergi. 1992.
Meet the Applegates. Dir. Michael Lehmann. Cinemarque/New World. 1990.
The Milagro Beanfield War. Dir. Robert Redford. Universal. 1988.
The Mission. Dir. Roland Joffe. Goldcrest/Kingsmore. 1986.
Mr. Kingstreet's War. Dir. Percival Rubens. HRS. 1973.
Moby-Dick. Dir. John Huston. Moulin. 1956.
The Mosquito Coast. Dir. Peter Weir. Warner Bros. 1986.
Namu, the Killer Whale. Dir. Lazslo Benedek. United Artists. 1966.
Never Cry Wolf. Dir. Carroll Ballard. Walt Disney Productions /Amarok. 1983.
On Deadly Ground. Dir. Steven Seagal. Warner Bros. 1994.
On the Beach. Dir. Stanley Kramer. Lomitas/United Artists. 1959.
Orca. Dir. Michael Anderson. Famous Films/Dino de Laurentiis. 1977.
Out of Africa. Dir. Sydney Pollack. Universal Pictures. 1985.
The Pelican Brief. Dir. Alan J. Pakula. Warner Bros. 1993.
Pocahontas. Dir. Michael Gabriel. Walt Disney Productions. 1995.
Red Dawn. Dir. John Milius. MGM. 1984.
Red River. Dir. Howard Hawks. Monterey/Charles K. Feldman Group. 1948.
Riders of the Whistling Pines. Dir. John English. Columbia. 1949.
The River. Dir. Mark Rydell. Universal. 1984.
A River Runs Through It. Dir. Robert Redford. Columbia/Allied Filmmakers. 1992.
The River Wild. Dir. Curtis Hanson. Turman-Foster/Universal. 1994.

Roar. Dir. Noel Marshall. Noel Marshall/Banjiro Uemura. 1981.
The Roots of Heaven. Dir. John Huston. 20th Century Fox/Darryl F. Zanuck. 1958.
The Savage Innocents. Dir. Nicholas Ray. Joseph Janni/Magic Film/Appia/Gray-Pathé. 1960.
The Sea Beast. Dir. Millard Webb. Warner Bros. 1926.
The Searchers. Dir. John Ford. Warner Bros. 1956.
Seven Days in May. Dir. John Frankenheimer. Joel/Seven Arts. 1964.
Shane. Dir. George Stevens. Paramount. 1953.
Silent Running. Dir. Douglas Trumball. Universal. 1971.
Silkwood. Dir. Mike Nichols. ABC/20th Century Fox. 1983.
The Snows of Kilimanjaro. Dir. Henry King. 20th Century Fox. 1952.
Soylent Green. Dir. Richard Fleischer. MGM. 1973.
Stagecoach. Dir. John Ford. Walter Wanger Productions. 1939.
Star Trek IV: The Voyage Home. Dir. Leonard Nimoy. Paramount. 1986.
Tarzan of the Apes. Dir. Sidney Scott. National Film Corporation of America. 1918.
Tarzan the Ape Man. Dir. W.S. Van Dyke. MGM. 1932.
Terminator 2: Judgement Day. Dir. James Cameron. Carolco/Pacific Western/Lightstorm/Le Studio Canal +. 1991.
Testament. Dir. Lynne Littman. Paramount. 1983.
Them! Dir. Gordon Douglas. Warner Bros. 1954.
The Thing. Dir. John Carpenter. Universal. 1982.
The Three Little Pigs. Dir. Burton Gillett. Walt Disney Productions. 1933.
Thunder Run. Dir. Gary Hudson. Thunder Films. 1985.
Thunderheart. Dir. Michael Apted. Columbia Tri-Star. 1992.
The Toxic Avenger. Dir. Michael Herz, Lloyd Kaufman. Troma. 1985.
Trader Horn. Dir. W.S. Van Dyke. MGM. 1931.
True Lies. Dir. James Cameron. Lightstorm/20th Century Fox/Universal. 1994.
Tucker: The Man and His Dream. Dir. Francis Ford Coppola. Lucasfilm. 1988.
The Two Jakes. Dir. Jack Nicholson. Paramount. 1990.
Valley of the Giants. Dir. Charles J. Brabin. First National. 1927.
Valley of the Giants. Dir. William Keighley. Warner Bros. 1938.
Wayne's World. Dir. Penelope Spheeris. Paramount. 1992.
The White Dawn. Dir. Philip Kaufman. Paramount. 1974.
White Fang. Dir. David Butler. 20th Century Fox. 1936.
White Fang. Dir. Randal Kleiser. Hybrid. 1991.
White Fang II: Myth of the White Wolf. Dir. Ken Olin. Walt Disney Productions. 1994.
White Hunter, Black Heart. Dir. Clint Eastwood. Warner Bros./Malpaso/Rastar. 1990.
Who Framed Roger Rabbit. Dir. Robert Zemeckis. Silverscreen/Touchstone/Amblin. 1988.
The Wild Bunch. Dir. Sam Peckinpah. Warner Bros./Seven Arts. 1969.
Wind Across the Everglades. Dir. Nicholas Ray. Warner Bros. 1958.

Bibliography

Adam, Barbara, *Timescapes of Modernity: The Environment and Invisible Hazards* (London and New York: Routledge, 1998)

Adams, Jonathan S. and Thomas O. McShane, *The Myth of Wild Africa: Conservation Without Illusion* (London: University of California Press, 1992)

Adamson, Joy, *Born Free: A Lioness of Two Worlds* (London: Fontana/Collins, 1960)

——, 'What Animals Can Teach Us', *Family Weekly*, 9 October 1966, 4

Aitken, Stuart C. and Leo E. Zonn, eds, *Place, Power, Situation, and Spectacle: A Geography of Film* (Lanham, MD: Rowman and Littlefield Publishers, Inc., 1994)

Aleiss, Angela, 'Le Bon Sauvage: *Dances With Wolves* and the Romantic Tradition', *American Indian Culture and Research Journal*, 15.4 (1991), 91–6

Allen, Paula Gunn, *The Sacred Hoop: Recovering the Feminine in American Indian Traditions* (Boston: Beacon Press, 1986)

Almendros, Nestor, 'Photographing "Days of Heaven"', *American Cinematographer*, June 1979, 562–632

Andrew, Geoff, *The Films of Nicholas Ray: The Poet at Nightfall* (London: Charles Letts, 1991)

Angus, Ian and Sut Jhally, eds, *Cultural Politics in Contemporary America* (London: Routledge, 1989)

Athanasiou, Tom, *Slow Reckoning: The Ecology of a Divided Planet* (London: Secker and Warburg, 1996)

Atlas, Jacoba, 'The Facts behind "Chinatown"', *Los Angeles Free Press*, 27 September 1974, 23

Babbit, Bruce, 'The Fierce Green Fire', *Audubon*, September/October 1994, 120

Baker, Steve, *Picturing the Beast: Animals, Identity and Representation* (Manchester and New York: Manchester University Press, 1993)

Bakhtin, Mikhail, *Rabelais and His World*, trans. Helene Iswolsky (Cambridge Mass.: MIT Press, 1984)

Barthes, Roland, *The Rustle of Language*, trans. Richard Howard (Oxford: Basil Blackwell, 1986)

Bazalgette, Cary and David Buckingham, eds, *In Front of the Children: Screen Entertainment and Young Audiences* (London: BFI Publishing, 1995)

Bell, Elizabeth, Lynda Haas and Laura Sells, eds, *From Mouse to Mermaid: The*

Politics of Film, Gender, and Culture (Bloomington and Indianapolis: Indiana University Press, 1995)

Benton, Ted, *Natural Relations: Ecology, Animal Rights and Social Justice* (London: Verso, 1993)

Berenstein, Rhoda J., *Attack of the Leading Ladies: Gender, Sexuality, and Spectatorship in Classic Horror Cinema* (New York: Columbia University Press, 1996)

Berger, John, *Ways of Seeing* (London: British Broadcasting Corporation, 1972)

Bergman, B.J., 'Earth in the Balcony', *Sierra*, November/December 1999, 51–68

Bergman, Charles, *Wild Echoes: Encounters with the Most Endangered Animals in North America* (Anchorage and Seattle: Alaska Northwest Books, 1990)

Berkman, Meredith, 'How "Dances" Got Real', *Entertainment*, 8 March 1991, 22

Berry, Thomas, *The Dream of the Earth* (San Francisco: Sierra Club Books, 1988)

Biehl, Janet, *Rethinking Ecofeminist Politics* (Boston: South End Press, 1991)

Blake, Michael, *Dances With Wolves* (London and New York: BCA, 1991)

Boime, Albert, *The Magisterial Gaze: Manifest Destiny and the American Landscape Painting, c. 1830–1865* (Washington DC: Smithsonian Institution, 1991)

Bordewich, Fergus M., *Killing the White Man's Indian: Reinventing Native Americans at the End of the Twentieth Century* (New York and London: Doubleday, 1996)

Bordwell, David, *On the History of Film Style* (Cambridge Mass. and London: Harvard University Press, 1997)

Bordwell, David and Noël Carroll, eds, *Post-Theory: Reconstructing Film Studies* (Madison and London: University of Wisconsin Press, 1996)

Bordwell, David, Janet Staiger and Kristin Thompson, *The Classical Hollywood Cinema: Film Style and Mode of Production to 1960* (New York: Columbia University Press, 1985)

Bordwell, David and Kristin Thompson, *Film Art: An Introduction* (New York: McGraw-Hill, Inc., 1979; 4th edn, 1993)

Botkin, Daniel B., *Discordant Harmonies: A New Ecology for the Twenty-First Century* (New York and Oxford: Oxford University Press, 1990)

Boyer, Paul, *By The Bomb's Early Light: American Thought and Culture at the Dawn of the Atomic Age* (New York: Pantheon Books, 1985)

Branch, Michael P., Rochelle Johnson, Daniel Patterson and Scott Slovic, eds, *Reading the Earth: New Directions in the Study of Literature and the Environment* (Moscow ID: University of Idaho Press, 1998)

Bright, Deborah, 'The Machine in the Garden Revisited: American Environmentalism and Photographic Aesthetics', *Art Journal*, 51.2 (Summer 1992), 60–71

Broderick, Mike, *Nuclear Movies: A Critical Analysis and Filmography* (Jefferson NC: McFarland, 1991)

Brookfield, Charles M. and Oliver Grisowld, *They All Called It Tropical: True Tales of the Romantic Everglades, Cape Sable, and the Florida Keys* (Miami: Historical Association of South Florida, 1949; 9th edn, 1985)

Browne, Nick, ed., *Refiguring American Film Genres: Theory and History* (Berkeley, Los Angeles and London: University of California Press, 1998)

Budiansky, Stephen, *Nature's Keepers: The New Science of Nature Management* (London: Phoenix Giant, 1996)

Burgoyne, Robert, *Film Nation: Hollywood Looks at U.S. History* (Minneapolis and London: University of Minnesota Press, 1997)
Burnham, David, 'Nuclear Experts Debate "The China Syndrome"', *New York Times*, 18 January 1979, 1ff.
Buscombe, Edward, ed., *The BFI Companion to the Western* (London: André Deutsch, 1988)
Cantrill, James G. and Christine L. Oravec, *The Symbolic Earth: Discourse and Our Creation of the Environment* (Lexington: University Press of Kentucky, 1996)
Care, Ross B., 'Threads of Melody—The Evolution of a Major Film Score—Walt Disney's *Bambi*', *Quarterly Journal of the Library of Congress*, Spring 1983, 79–98
Carlton, Michael, 'Disney Environmentalists', *Southern Living*, 30.12 December 1995, 90–4
Carroll, Noël, *Mystifying Movies: Fads and Fallacies in Contemporary Film Theory* (New York: Columbia University Press, 1988)
——, *Theorizing the Moving Image* (Cambridge: Cambridge University Press, 1996)
Carson, Rachel, *Silent Spring* (1962) (Harmondsworth: Penguin, 1991)
Cartmill, Matt, *A View to a Death in the Morning: Hunting and Nature Through History* (London and Cambridge Mass.: Harvard University Press, 1993)
Christensen, Terry, *Reel Politics: American Political Movies from 'Birth of a Nation' to 'Platoon'* (Oxford and New York: Basil Blackwell, 1987)
Clarfield, Gerald H. and William M. Wiecek, *Nuclear America: Military and Civil Nuclear Power in the United States 1940–1980* (New York: Harper and Row, 1984)
Clarke, David B., ed., *The Cinematic City* (London and New York: Routledge, 1997)
Clover, Carol, *Men, Women and Chainsaws: Gender in the Modern Horror Film* (London: BFI Publishing, 1992)
Coates, Peter, *The Trans-Alaska Pipeline Controversy* (London: Bethlehem Lehigh University Press, 1991)
——, *In Nature's Defence: Americans and Conservation* (Keele: British Association for American Studies, 1993)
——, *Nature: Western Attitudes since Ancient Times* (Berkeley, Los Angeles and London: University of California Press, 1998)
Cochrane, Willard W., *The Development of American Agriculture: A Historical Analysis* (Minneapolis and London: University of Minnesota Press, 1993)
Cohan, Steven and Ina Rae Hark, eds, *The Road Movie Book* (London and New York: Routledge, 1997)
Commoner, Barry, *The Closing Circle: Nature, Man and Technology* (New York: Alfred A. Knopf, 1972)
Conley, Verena Andermatt, *Ecopolitics: The Environment in Poststructuralist Thought* (London and New York: Routledge, 1997)
Cosgrove, Denis, *Social Formation and Symbolic Landscape* (London and Sydney: Croom Helm, 1984)
Cousteau, Jacques-Yves and the staff of the Cousteau Society, *The Cousteau Almanac: An Inventory of Life on Our Water Planet* (New York: Doubleday, 1981)
Creed, Barbara, *The Monstrous Feminine: Film, Feminism, Psychoanalysis* (London and New York: Routledge, 1993)

Cronon, William, ed., *Uncommon Ground: Rethinking the Human Place in Nature* (New York and London: W.W. Norton and Company, 1996)
Crowther, Bosley, 'The Wolf Becomes a Hero', *New York Times*, 10 November 1962, 16
Custen, George F., *Bio/Pics: How Hollywood Constructed Public History* (New Brunswick: Rutgers University Press, 1992)
Davis, Erik, review of *FernGully*, *Village Voice*, 22 April 1992, 63
Davis, Ronald, 'Paradise Among the Monuments: John Ford's Vision of the American West', *Montana: The Magazine of Western History*, Summer 1995, 48–63
Davis, Susan G., *Spectacular Nature: Corporate Culture and the Sea World Experience* (Berkeley and Los Angeles: University of California Press, 1997)
Del Mar, David Peterson, '"Our Animal Friends": Depictions of Animals in *Reader's Digest* During the 1950s', *Environmental History*, 3.1 (January 1998), 25–44
Delehanty, Thomas, 'The New Film', *Post*, 2 April 1931
Deleuze, Gilles and Félix Guattari, *A Thousand Plateaus: Capitalism and Schizophrenia*, trans. Brian Massumi (London: Athlone Press, 1988)
Dempsey, Michael, 'Quatsi Means Life: The Films of Godfrey Reggio', *Film Quarterly*, Spring 1989, 2–12
Denvir, John, ed., *Legal Reelism: Movies as Legal Texts* (Urbana and Chicago: University of Illinois Press, 1996)
Devall, Bill and George Sessions, *Deep Ecology: Living as if Nature Mattered* (Salt Lake City: Gibbs Smith, 1985)
Diggs, Terry, 'Welcome to the All-Republican Cineplex', *Legal Times*, 19 June 1995, 58
Drummond, Lee, *American Dreamtime: A Cultural Analysis of Popular Movies, and their Implications for a Science of Humanity* (London: Littlefield Adams, 1996)
Dunlap, Thomas R., *Saving America's Wildlife* (New Jersey: Princeton University Press, 1988)
Easlea, Brian, *Fathering the Unthinkable: Masculinity, Scientists and the Nuclear Arms Race* (London: Pluto Press, 1983)
Easterbrook, Gregg, *A Moment on the Earth: The Coming Age of Environmental Optimism* (Harmondsworth: Penguin, 1995)
Easton, Nina J., 'Film Makers in the Mist', *Los Angeles Times*, 16 September 1988, 28
Eaton, Michael, *Chinatown* (London: BFI Publishing 1997)
Egan, Philip S., *Design and Destiny: The Making of the Tucker Automobile* (Orange: On the Mark, 1989)
Ehrlich, Paul, *The Population Bomb* (New York: Ballantine Books, 1968)
Eisenschitz, Bernard, *Nicholas Ray: An American Journey* (London and Boston: Faber and Faber, 1993)
Eisner, Michael, *Work in Progress* (New York: Random House, 1998)
Evans, Joyce A., *Celluloid Mushroom Clouds: Hollywood and the Atomic Bomb* (Boulder and Oxford: Westview, 1998)
Evernden, Neil, *The Social Creation of Nature* (Baltimore and London: Johns Hopkins University Press, 1992)
——, *The Natural Alien* (Toronto, Buffalo and London: University of Toronto Press,

1985; 2nd edn, 1993)
Fleming, Anne Taylor, 'Turning Stars Into Environmentalists', *New York Times*, 25 October 1989, C8
Flynn, Hazel 'You'll Laugh, Cry Over Wolf Hero', *Citizen-News*, 10 November 1962
Fogelman, Valerie, 'American Attitudes Towards Wolves: A History of Misperception', *Environmental Review*, 13.1 (Spring 1989), 63–94
Foreman, Carl, 'Foreman the Lion Tamer Extolls On Bearding the Beast on Film', *Variety Weekly*, 13 January 1965, 22
Fossey, Dian, *Gorillas in the Mist* (Harmondsworth: Penguin, 1983)
Foucault, Michel, *Discipline and Punish: The Birth of the Prison*, trans. Alan Sheridan (London: Allen Lane, 1977)
Galbraith, Jane, 'A Whale of an Actor in a Killer Part', *Los Angeles Times*, 16 May 1993, 23
Gall, Sandy, *George Adamson: Lord of the Lions* (London: Grafton Books, 1991)
Gare, Arran E., *Postmodernism and the Environmental Crisis* (London and New York: Routledge, 1995)
Gary, Romain, *The Roots of Heaven* (London: Michael Joseph, 1958)
George, Terry, 'Hollywood Goes Green', *Audubon*, March/April 1992, 86
Giago, Tim, Mary Cook, and Gemma Lockhart, 'They've Gotten It Right This Time', *Native Peoples*, Winter 1991, 6–14
Giblett, Rod, *Postmodern Wetlands: Culture, History, Ecology* (Edinburgh: Edinburgh University Press, 1996)
Gillette, Robert, '"China Syndrome": Nuclear Reactions', *Los Angeles Times*, 25 March 1979, 3
Goodman, Joan, 'How Hollywood Whispers in the President's Ear', *Sunday Times Magazine*, 23 March 1997, 20–24
Gould, Stephen Jay, *The Panda's Thumb* (Harmondsworth: Penguin, 1980)
Granberry, Michael, 'Free Corky?', *Los Angeles Times*, 11 August 1993, A3
Grant, Barry K., ed., *Film Genre Reader II* (Austin: University of Texas Press, 1995)
Gross, Paul R. and Norman Levitt, *Higher Superstition: The Academic Left and Its Quarrels with Science* (Baltimore: Johns Hopkins University Press, 1994)
Gross, Paul R., Norman Levitt, and Martin W. Lewis, eds, *The Flight From Science and Reason* (New York: New York Academy of Sciences, 1996)
Grossberg, Lawrence, Cary Nelson and Paula A. Treichler, eds, *Cultural Studies* (New York and London: Routledge, 1992)
Grumbine, R. Edward, *Ghost Bears: Exploring the Biodiversity Crisis* (Washington DC and Covelo: Island Press, 1992)
Hanson, Victor Davis, *Fields Without Dreams: Defending the Agrarian Idea* (New York and London: The Free Press, 1996)
Haraway, Donna, *Primate Visions: Gender, Race, and Nature in the World of Modern Science* (London and New York: Verso, 1989)
——, *Simians, Cyborgs, and Women: The Reinvention of Nature* (New York: Routledge, 1991)
Harr, Jonathan, *A Civil Action* (New York: Vintage Books, 1995)
Harvey, David, *Justice, Nature and the Geography of Difference* (Oxford and Cambridge

Mass.: Blackwell, 1996)
Hayes, Harold T.P., *The Dark Romance of Dian Fossey* (New York: Simon and Schuster, 1990)
Hays, Samuel P., *Beauty, Health, and Permanence: Environmental Politics in the United States, 1955–1985* (Cambridge, New York etc.: Cambridge University Press, 1987)
Heath, Stephen, 'Jaws, Ideology and Film Theory', *Framework*, 4 (Autumn 1976), 25–7
Heise, Ursula K., 'What is Ecocriticism?', Association for the Study of Literature and the Environment web-site, at http://www.asle.umn.edu/
Hendricks, Gordon, *Albert Bierstadt: Painter of the American West* (New York: Harry N. Abrams, Inc., 1974)
Henley, Don and Dave Marsh, *Heaven is Under Our Feet: A Book For Walden Woods* (New York: Berkley Books, 1991)
Herndl, Carl G. and Stuart C. Brown, eds, *Green Culture: Environmental Rhetoric in Contemporary America* (Madison and London: University of Wisconsin Press, 1996)
Hillier, Jim, *The New Hollywood* (London: Studio Vista, 1992)
Hillier, Jim and Peter Wollen, eds, *Howard Hawks: American Artist* (London: BFI Publishing, 1996)
Hochman, Jhan, *Green Cultural Studies: Nature in Film, Novel, and Theory* (Moscow ID: University of Idaho Press,1998)
Hofmeister, Sallie, 'In the Realm of Marketing, the "Lion King" Rules', *New York Times*, 12 July 1994, D17
House, Adrian, *The Great Safari: The Lives of George and Joy Adamson* (London: Harvill, 1993)
Hunter, Stephen, '*Never Cry Wolf* Offers Viewers the Spectacular Wonder of Nature', *Baltimore Sun*, 11 November 1983, B4
Huston, John, *An Open Book* (New York: Alfred A. Knopf, 1980)
Huth, Hans, *Nature and the American: Three Centuries of Changing Attitudes* (Berkeley and Los Angeles: University of California Press, 1957)
Jagtenberg, Tom and David McKie, *Eco-Impacts and the Greening of Modernity* (London: Sage, 1997)
Jameson, Fredric, *Postmodernism, or, The Cultural Logic of Late Capitalism* (London and New York: Verso, 1991)
——, *The Geopolitical Aesthetic: Cinema and Space in the World System* (London: BFI Publishing, 1992)
Jay, Martin, *Downcast Eyes: The Denigration of Vision in Twentieth-Century French Thought* (Berkeley and London: University of California Press, 1994)
Jewel, Dan and Mark Dagostino, 'A Civil Warrior', *People*, 2 August 1999, 79–82
Jewett, Robert and John Shelton Lawrence, *The American Monomyth* (Lanham MD: University Press of America, 1988)
Kasdan, Margo and Susan Tavernetti, 'The Hollywood Indian in *Little Big Man*: A Revisionist View', *Film and History* XXIII, 1.4 (1993), 75–6
Kearney, Jill, 'The Old Gringo', *American Film*, March 1988, 67
Kennedy, Dan, 'Take Two', *Boston Phoenix*, 2 January 1998, 14–16

Kerasote, Ted, 'Disney's New Nature Myth', *Audubon*, November/December 1994, 132.
Kerridge, Richard and Neil Sammells, eds, *Writing the Environment: Ecocriticism and Literature* (London and New York: Zed Book Ltd, 1998)
Kirkham, Pat and Janet Thumin, eds, *You Tarzan: Masculinity, Movies and Men* (London: Lawrence and Wishart Ltd, 1993)
Kolodny, Annette, *The Land Before Her: Fantasy and Experience of the American Frontiers, 1630–1860* (Chapel Hill and London: University of North Carolina Press, 1984)
Kopecky, Gini, '"Free Willy" . . . and Maybe Rescue Keiko, Too', *New York Times*, 11 July 1993, 10
Krech III, Shepard, *The Ecological Indian: Myth and History* (New York and London: W.W. Norton and Company, 1999)
Kuhn, Annette, ed., *Alien Zone: Cultural Theory and Contemporary Science Fiction* (London: Verso, 1990)
Kyne, Peter B., *The Valley of the Giants* (New York: Grosset and Dunlap, 1918)
Laermer, Richard, 'Cars Don't Have to Take a Back Seat in the Movies', *New York Times*, 20 November 1988, H21
Lake, Randall A., 'Argumentation and Self: The Enactment of Identity in *Dances With Wolves*', *Argumentation and Advocacy*, 34 (Fall 1997), 66–89
Lansing, Robert, 'Namu: Nice Guy Killer Whale', *Los Angeles Times*, Calendar, 24 July 1966
Lash, Scott, Bronislaw Szerszynski and Brian Wynne, *Risk, Environment and Modernity: Towards a New Ecology* (London, Thousand Oaks, New Delhi: Sage, 1996)
Latham, Aaron, 'Hollywood vs. Harrisburg', *Esquire*, 22 May 1979, 77–86
Lefebvre, Henri, *The Production of Space*, trans. Donald Nicholson-Smith (1974) (Oxford and Cambridge Mass.: Blackwell, 1991)
Leonelli, Elisa, 'Hector Babenco', *Venice Magazine*, December 1991, 34–5
Leslie, Robert Frank, *The Bears and I: Raising Three Cubs in the North Woods* (New York: Ballantine, 1968)
Lester, J.P., ed., *Environmental Politics and Policy: Theories and Evidence* (Durham and London: Duke University Press, 1989)
Leuthold, Steven M., 'Native American Responses to the Western', *American Indian Culture and Research Journal*, 19.1 (1995), 181–5
Lewis, Martin W., *Green Delusions: An Environmentalist Critique of Radical Environmentalism* (Durham and London: Duke University Press, 1992)
Lieberman, Jane, 'Hollywood Who's Who Backs EMA', *Variety*, 10 August 1992, 50
Limerick, Patricia Nelson, *The Legacy of Conquest: The Unbroken Past of the American West* (New York and London: Norton, 1987)
London, Jack, *The Call of the Wild, White Fang, and Other Stories* (Harmondsworth: Penguin, 1981)
Long, William J., *Mother Nature: A Study of Animal Life and Death* (New York and London: Harper and Bros., 1923)
Lopez, Barry Holstun, *Of Wolves and Men* (New York: Touchstone, 1978)
Love, Glen, 'Science, Anti-Science, and Ecocriticism', *ISLE: Interdisciplinary Studies in Literature and the Environment*, 6.1 (Winter 1999), 65–81
Lovell, Terry, *Pictures of Reality: Aesthetics, Politics and Pleasure* (London: BFI

Publishing, 1983)
Lutts, Ralph H., *The Nature Fakers: Wildlife, Science and Sentiment* (Golden CO: Fulcrum Publishing, 1990)
——, 'The Trouble With Bambi: Walt Disney's *Bambi* and the American Vision of Nature', *Forest and Conservation History*, 36.4 (October 1992), 160–71
——, 'Chemical Fallout and the Environmental Movement', in Allan M. Winkler, ed., *The Recent Past: Readings on America Since World War Two* (New York: Harper and Row, 1989)
——, ed., *The Wild Animal Story* (Philadelphia: Temple University Press, 1998)
McBride, Joseph, *Hawks on Hawks* (London: University of California Press, 1982)
McIntyre, Rick, ed., *War Against the Wolf: America's Campaign to Exterminate the Wolf* (Stillwater MN: Voyager Press, 1995)
McKibben, Bill, *The End of Nature* (New York: Random House, 1989)
——, *The Age of Missing Information* (New York: Random House, 1992)
Maclean, Norman, *A River Runs Through It, and Other Stories* (Chicago and London: University of Chicago Press, 1976)
Madsen, Axel, *John Huston* (London: Robson Books, 1979)
Maltby, Richard and Ian Craven, *Hollywood Cinema* (Oxford and Cambridge Mass.: Blackwell, 1995)
Martin, Phillip W.D., 'Quiet, Please, On the Set: Birds Nesting', *New York Times*, 19 February 1995, 13–14
Matthiessen, Peter, *In the Spirit of Crazy Horse* (New York: Collins-Harvill, 1992)
Melville, Herman, *Moby-Dick, or, The Whale* (Harmondsworth: Penguin, 1972)
Merchant, Carolyn, *Radical Ecology: The Search for a Livable World* (New York and London: Routledge, 1992)
Mighetto, Lisa, *Wild Animals and Environmental Ethics* (Tucson: University of Arizona Press, 1991)
Miller, George, 'Progress for Indians is a Film Fantasy', *Los Angeles Times*, 26 March 1991, B7
Mitman, Gregg, *Reel Nature: America's Romance with Wildlife on Film* (Cambridge Mass.: and London: Harvard University Press, 1999)
Modleski, Tania, *Feminism Without Women: Culture and Criticism in a "Post-feminist" Age* (New York and London: Routledge, 1991)
Morone, Joseph G. and Edward J. Woodhouse, *The Demise of Nuclear Energy? Lessons for Democratic Control of Technology* (New Haven and London: Yale University Press, 1989)
Motavelli, Jim and Susan Elan, 'Jacques Yves Cousteau at 85', *E Magazine*, March/April 1996, 10–12
Mottram, Eric, *Blood on the Nash Ambassador: Investigations in American Culture* (London: Hutchinson Radius, 1983)
Mowat, Farley, *Never Cry Wolf* (New York: Laurel, 1963)
Mullan, Bob and Garry Marvin, *Zoo Culture* (London: Weidenfeld and Nicolson, 1987)
Nash, Roderick, 'The Exporting and Importing of Nature: Nature Appreciation as a Commodity 1850–1980', *Perspectives in American History*, XII (1979), 521–60
——, *Wilderness and the American Mind* (London: Yale University Press, 1982)

Neale, Steven, 'The Same Old Story: Stereotypes and Difference', *Screen Education*, 32–3 (Autumn/Winter 1979–80), 33–7

Newman, Anthony, 'Steven Seagal's Box-Office Smash Scorned in Alaska', *Los Angeles Times*, 4 March 1996, F6

Newman, Kim, *Millennium Movies: End of the World Cinema* (London: Titan, 1999)

Nichols, Bill, ed., *Movies and Methods Vol. II* (Berkeley and Los Angeles: University of California Press, 1985)

Nichols, John, 'Bob and the Bean Stalk', *American Film*, May 1987, 13–17

Norris, Christopher, *Uncritical Theory: Postmodernism, Intellectuals and the Gulf War* (London: Lawrence and Winshart, 1992)

——, *Against Relativism: Philosophy of Science, Deconstruction and Critical Theory* (Oxford: Blackwell, 1997)

Norwood, Vera, *Made From the Earth: American Women and Nature* (Chapel Hill: University of North Carolina Press, 1993)

Noske, Barbara, *Humans and Other Animals* (London: Pluto, 1989)

Novak, Barbara, *Nature and Culture: American Landscape Painting, 1825–1875* (New York: Oxford University Press, 1995)

Nye, David, *Narratives and Spaces: Technology and the Construction of American Culture* (Exeter: University of Exeter Press, 1997)

O'Brien, Tom, *The Screening of America* (New York: Continuum, 1990)

Oelschlaeger, Max, *The Idea of Wilderness: From Prehistory to the Age of Ecology* (New Haven and London: Yale University Press, 1991)

Okrent, Neil, '*At Play in the Fields of the Lord:* An Interview With Hector Babenco', *Cineaste*, 19.1 (1992), 44–7

Ornstein, Robert and Paul Ehrlich, *New World, New Mind: Moving Toward Conscious Evolution* (New York: Simon and Schuster, 1989)

Padget, Martin, 'Film, Ethnography and the Scene of History: *Dances With Wolves* and Participant Observation', *Borderlines* 3.4 (1996), 396–412

Palmquist, Peter E., *Carelton E. Watkins: Photographer of the American West* (Albuquerque: University of New Mexico Press, 1983)

Paterson, Richard, 'The River', *American Cinematographer*, November 1984, 66–72

Patterson, John, 'Sequel Opportunities, *Neon*, August 1998, 24

Payne, Roger, *Among Whales* (New York: Scribner, 1995)

Pepper, David, *Eco-Socialism: From Deep Ecology to Social Justice* (London and New York: Routledge, 1993)

Perrine, Toni, *Film and the Nuclear Age: Representing Nuclear Anxiety* (New York and London: Garland, 1998)

Peterson, John M., *Aim on Target: The FBI's War on Leonard Peltier and the American Indian Movement* (John M. Peterson, 1994)

Pfeil, Fred, *White Guys: Studies in Postmodern Domination and Difference* (London: Verso, 1995)

Polanski, Roman, 'The Day I Gave Jack Nicholson a Bloody Nose', *Gentleman's Quarterly*, February 1984, 178–215

Powers, James, 'Disney Production Holds Wide Appeal', *Hollywood Reporter*, 24 October 1962, 3

Prince, Stephen, *Visions of Empire: Political Imagery in Contemporary American Film*

(New York: Praeger, 1992)

Pristin, Terry, 'The Filmmakers vs. The Crusaders', *Los Angeles Times*, Calendar, 29 December 1991, 7–33

Pyne, Stephen, *Fire in America: A Cultural History of Wildland and Rural Fire* (Princeton NJ: Princeton University Press, 1982)

Ray, Robert, *A Certain Tendency of the Hollywood Cinema, 1930–1980* (Princeton NJ: Princeton University Press, 1985)

Regan, Tom, *All That Dwells Therein: Animal Rights and Environmental Ethics* (Berkeley: University of California Press, 1982)

Regenstein, Lewis, *The Politics of Extinction: The Shocking Story of the World's Endangered Wildlife* (New York and London: Macmillan, 1975)

Reisner, Marc, *Cadillac Desert: The American West and Its Disappearing Water* (London: Secker & Warburg, 1990)

Roberts, Gerrylynn K. and Philip Steadman, *American Cities and Technology: Wilderness to Wired City* (London and New York: Routledge, 1999)

Robertson, George, Melinda Mash, Lisa Tickner, Jon Bird, Barry Curtis and Tim Putnam, eds, *FutureNatural: Nature, Science, Culture* (London and New York: Routledge, 1996)

Rodowick, D.N., *The Crisis of Political Modernism: Criticism and Ideology in Contemporary Film Theory* (Urbana and Chicago: University of Illinois Press, 1999)

Rollins, Peter C. and John E. O'Connor, eds, *Hollywood's Indian: The Portrayal of the Native American in Film* (Lexington: University Press of Kentucky, 1998)

Rony, Fatimah Tobing, *The Third Eye: Race, Cinema and Ethnographic Spectacle* (Durham and London: Duke University Press, 1996)

Rosen, Philip, ed., *Narrative, Apparatus, Ideology: A Film Theory Reader* (New York: Columbia University Press, 1986)

Ross, Andrew, *Strange Weather: Culture, Science and Technology in the Age of Limits* (London: Verso, 1991)

——, *The Chicago Gangster Theory of Life* (London and New York: Verso, 1994)

——, ed., *No Sweat: Fashion, Free Trade, and the Rights of Garment Workers* (New York and London: Verso, 1997)

Runte, Alfred, *National Parks: The American Experience* (Lincoln: University of Nebraska Press, 1979)

Ryan, Michael and Douglas Kellner, *Camera Politica: The Politics and Ideology of Contemporary Hollywood Film* (Bloomington and Indianapolis: Indiana University Press, 1990)

Sale, Kirkpatrick, *Dwellers in the Land: The Bioregional Vision* (San Francisco: Sierra Club Books, 1985)

Salleh, Ariel, *Ecofeminism as Politics: Nature, Marx and the Postmodern* (London and New York: Zed Books, 1997)

Sarf, Wayne Michael, 'Oscar Eaten by Wolves', *Film Comment*, November/December 1991, 62–70

Schaefer, Stephen, 'Bored With His Name', interview with Robert Redford, *Film Comment*, February 1988, 36–41

Schama, Simon, *Landscape and Memory* (London: HarperCollins, 1995)

——, 'The Princess of Eco-Kitsch', *New York Times*, 14 June 1995, A21

Scheffer, Victor B., *The Shaping of Environmentalism in America* (Seattle and London: University of Washington Press, 1991)

Schickel, Richard, *The Disney Version* (London: Pavilion, 1986)

Schlickeisen, Rodger, 'Wolf Recovery's Significance', *Half a Hundred: Defenders Magazine*, Winter 1996/7, on Defenders of Wildlife web-site, www.defenders.org, 18 August 1997

Schmitt, Peter J., *Back to Nature: The Arcadian Myth in Urban America* (Baltimore and London: Johns Hopkins University Press, 1990)

Schrepfer, Susan, *The Fight to Save the Redwoods: A History of Environmental Reform 1917–1978* (Wisconsin and London: University of Wisconsin Press, 1983)

Schulberg, Budd, *Across the Everglades: A Play for the Screen* (New York: Random House, 1958)

——, with photographs by Geraldine Brooks, *Swan Watch* (London: Robson Books, 1975)

Schullery, Paul, '*Bambi* and the Fires of Yellowstone', *Backpacker*, December 1990, 95–6

Schwarz, O. Douglas, 'Indian Rights and Environmental Ethics', *Environmental Ethics*, 9.4 (Winter 1989), 295–8

Seligman, David, 'Talking Back to Trees', *Fortune*, 29 June 1992, 2

Seton, Ernest Thompson, *Wild Animals I Have Known* (1898) (Harmondsworth: Penguin, 1987)

Shaheen, Jack, *Nuclear War Films* (Carbondale and London: Southern Illinois University, 1978)

Shepard, Paul and Barry Sanders, *The Sacred Paw: The Bear in Nature, Myth, and Literature* (New York: Viking, 1985)

Sherow, James E., 'Workings of the Geodialectic: High Plains Indians and Their Horses in the Regions of the Arkansas River Valley, 1800–1870', *Environmental History Review*, 16.2 (Summer 1992), 61–84

Shohat, Ella and Robert Stam, *Unthinking Eurocentrism: Multiculturalism and the Media* (London and New York: Routledge, 1994)

Slotkin, Richard, *Gunfighter Nation: The Myth of the Frontier in Twentieth-Century America* (New York: Atheneum, 1992)

Smith, Henry Nash, *Virgin Land* (Cambridge Mass.: Harvard University Press, 1950)

Smith, Murray, *Engaging Characters: Fiction, Emotion and the Cinema* (Oxford: Oxford University Press, 1995)

Smith, Neil, *Uneven Development: Nature, Capital and the Production of Space* (Oxford and Cambridge Mass.: Basil Blackwell, 1990)

Sobchack, Vivian, ed., *The Persistence of History: Cinema, Television and the Modern Event* (New York: Routledge, 1996)

Sokal, Alan and Jean Bricmont, *Fashionable Nonsense: Postmodern Intellectuals' Abuse of Science (*New York: Picador, 1998)

Solomon, J. Fisher, *Discourse and Reference in the Nuclear Age* (Norman : University of Oklahoma Press, 1988)

Sontag, Susan, *On Photography* (Harmondsworth: Penguin, 1977)

Soper, Kate, *What is Nature? Culture, Politics and the Non-Human* (Oxford and Cambridge Mass.: Blackwell, 1995)

Spence, Mark David, *Dispossessing the Wilderness: Indian Removal and the Making of National Parks* (New York and Oxford: Oxford University Press, 1999)

Starhawk, *The Spiral Dance: A Rebirth of the Ancient Religion of the Great Goddess* (San Francisco and London: Harper and Row, 1979)

Stauth, Cameron, 'Eco Trip: Hollywood Turns Green-eyed in a Multipicture Deal with Planet Earth', *American Film*, November 1990, 16

Steiner, Stan, *The Vanishing White Man* (Norman and London: University of Oklahoma Press, 1987)

Sterling, John, 'The World According to Disney', *Earth Island Journal*, Summer 1994, 32–3

Sullivan, George, *Not Guilty: Five Times When Justice Failed* (New York and London: Scholastic Inc., 1997)

Swan, James A., *In Defense of Hunting* (New York: Harper Collins, 1995)

Tasker, Yvonne, *Spectacular Bodies: Gender, Genre and the Action Cinema* (London and New York: Routledge, 1993)

Tilton, Robert, *Pocahontas: The Evolution of an American Narrative* (Cambridge: Cambridge University Press, 1994)

Tompkins, Jane, *West of Everything: The Inner Life of Westerns* (New York and Oxford: Oxford University Press, 1992)

Toplin, Robert Brent, *History by Hollywood: The Use and Abuse of the American Past* (Chicago: University of Illinois Press, 1996)

Tors, Ivan, *My Life in the Wild* (Boston: Houghton Mifflin Co., 1979)

Towne, Robert, 'It's only LA, Jake', *Los Angeles Times*, 29 May 1994, 1–12

Truettner, William H., ed., *The West as America: Reinterpreting Images of the Frontier, 1820–1920* (Washington and London: Smithsonian Institution Press, 1991)

Tuan, Yi-Fu, *Passing Strange and Wonderful: Aesthetics, Nature, and Culture* (Washington DC and Covelo: Island Press/Shearwater Books, 1993)

Tunney, Tom, review of *On Deadly Ground*, *Sight and Sound*, May 1994, 48

Turner, Frederick Jackson, *The Frontier in American History* (New York: Henry Holt, 1920)

Tyler, David, 'Real Story is Lost in Movie Version of 'A Civil Action', *Daily Transcript* (Dedham, Massachusetts), 6 January 1999, 9

Urry, John, *The Tourist Gaze: Leisure and Travel in Contemporary Societies* (London: Sage, 1992)

Valente, Judith, 'A Century Later, Sioux Still Struggle, And Still Are Losing', *Wall Street Journal*, 25 March 1991, A1–2

Verschaffel, Bart and Mark Verminck, eds, *Zoology: On (Post) Modern Animals* (Dublin: The Lilliput Press; Leuven/Amsterdam: Kritak, 1993)

Weart, Spencer, *Nuclear Fear: A History of Images* (Cambridge Mass. and London: Harvard University Press, 1988)

Webster, Duncan, *Looka Yonder: The Imaginary America of Populist Culture* (London: Comedia, 1988)

Weinraub, Bernard, 'Good Guys, Bad Guys', *International Herald Tribune*, 13 January 1999, 9

Weisser, Pete, 'Hunting in the Movies', *Outdoor California*, November/December 1989, 1–7

Wicker, Tom, 'Waiting for an Environmental President', *Audubon*, September/October 1994, 49–103

Wilford, John Noble, 'A New Movie Boasts Unlikely Stars—Wolves', *New York Times*, 9 October 1983, 21

Williams, Christopher, ed., *Realism and the Cinema: A Reader* (London: Routledge & Kegan Paul, 1980)

Williams, Linda Ruth, 'Blood Brothers', *Sight and Sound*, September 1994, 16–19

Williams, Raymond, *The Country and the City* (New York: Oxford University Press, 1973)

Williamson, J.W., *Hillbillyland: What the Movies Did to the Mountains and What the Mountains Did to the Movies* (Chapel Hill and London: University of North Carolina Press, 1995)

Wilson, Alexander, *The Culture of Nature: North American Landscape from Disney to the Exxon Valdez* (Oxford: Blackwell, 1992)

Wilson, John M., 'Widening Film's Culture', *Writer's Digest*, March 1987, 48

Wood, Michael, *America in the Movies, or "Santa Maria, It Had Slipped My Mind"* (New York: Columbia University Press, 1975)

Worster, Donald, ed., *The Ends of the Earth: Perspectives on Modern Environmental History* (Cambridge: Cambridge University Press, 1988)

——, *Nature's Economy: A History of Ecological Ideas* (Cambridge: Cambridge University Press, 1994)

——, *Under Western Skies: Nature and History in the American West* (New York and Oxford: Oxford University Press, 1992)

Wrathall, John, review of *The American President*, *Sight and Sound*, January 1996, 36

Wright, Denis, 'Indian chief slams medical provision on Babenco film', *Screen International*, March 15–21 1991, 7

Zizek, Slavoj, *Looking Awry: An Introduction to Jacques Lacan through Popular Culture* (Cambridge Mass. and London: MIT Press, 1991)

Index

CPSIA information can be obtained
at www.ICGtesting.com
Printed in the USA
LVHW011711190121
676895LV00013B/1940

9 780859 896092